Electrochemical Rehabilitation Methods for Reinforced Concrete Structures – A State of the Art Report

T0133880

EARLIER VOLUMES IN THIS SERIES

European Federation of Corrosion Publications
Publications
NUMBER 24

Electrochemical Rehabilitation Methods for Reinforced Concrete Structures

A State of the Art Report

Edited by
J. MIETZ

Published for the European Federation of Corrosion
by The Institute of Materials

Book Number 709
Published in 1998 by IOM Communications Ltd
1 Carlton House Terrace, London SW1Y 5DB

IOM Communications Ltd
is a wholly-owned subsidiary of
The Institute of Materials

ISBN 1-86125-082-7

Design and production by
SPIRES Design Partnership

Made and printed in Great Britain

Contents

European Federation of Corrosion Publications
Series Introduction

The EFC, incorporated in Belgium, was founded in 1955 with the purpose of promoting European co-operation in the fields of research into corrosion and corrosion prevention.

Membership is based upon participation by corrosion societies and committees in technical Working Parties. Member societies appoint delegates to Working Parties, whose membership is expanded by personal corresponding membership.

The activities of the Working Parties cover corrosion topics associated with inhibition, education, reinforcement in concrete, microbial effects, hot gases and combustion products, environment sensitive fracture, marine environments, surface science, physico–chemical methods of measurement, the nuclear industry, computer based information systems, the oil and gas industry, the petrochemical industry and coatings. Working Parties on other topics are established as required.

The Working Parties function in various ways, e.g. by preparing reports, organising symposia, conducting intensive courses and producing instructional material, including films. The activities of the Working Parties are co-ordinated, through a Science and Technology Advisory Committee, by the Scientific Secretary.

The administration of the EFC is handled by three Secretariats: DECHEMA e. V. in Germany, the Société de Chimie Industrielle in France, and The Institute of Materials in the United Kingdom. These three Secretariats meet at the Board of Administrators of the EFC. There is an annual General Assembly at which delegates from all member societies meet to determine and approve EFC policy. News of EFC activities, forthcoming conferences, courses etc. is published in a range of accredited corrosion and certain other journals throughout Europe. More detailed descriptions of activities are given in a Newsletter prepared by the Scientific Secretary.

The output of the EFC takes various forms. Papers on particular topics, for example, reviews or results of experimental work, may be published in scientific and technical journals in one or more countries in Europe. Conference proceedings are often published by the organisation responsible for the conference.

In 1987 the, then, Institute of Metals was appointed as the official EFC publisher. Although the arrangement is non-exclusive and other routes for publication are still available, it is expected that the Working Parties of the EFC will use The Institute of Materials for publication of reports, proceedings etc. wherever possible.

The name of The Institute of Metals was changed to The Institute of Materials with effect from 1 January 1992.

A. D. Mercer
EFC Series Editor,
The Institute of Materials, London, UK

EFC Secretariats are located at:

Dr B A Rickinson
European Federation of Corrosion, The Institute of Materials, 1 Carlton House Terrace, London, SW1Y 5DB, UK

Mr P Berge
Fédération Européene de la Corrosion, Société de Chimie Industrielle, 28 rue Saint-Dominique, F-75007 Paris, FRANCE

Professor Dr G Kreysa
Europäische Föderation Korrosion, DECHEMA e. V., Theodor-Heuss-Allee 25, D-60486, Frankfurt, GERMANY

Preface

The corrosion of steel reinforcement in concrete is a serious problem on a worldwide scale. The European Federation of Corrosion (EFC) has addressed this problem by forming a Working Party related to the subject. Like other Working Parties of the EFC the main ambition of this Working Party is the transfer of knowledge from research work into practical application and vice versa. Recently the Working Party on Corrosion of Reinforcement in Concrete prepared a review on the use of stainless steel in concrete and this has been published in the EFC series as EFC 18. Another topic on the list of work items of the Working Party is concerned with electrochemical rehabilitation methods for steel reinforced concrete structures, e.g. cathodic protection, realkalisation and chloride extraction.

Cathodic Protection of reinforced concrete has been practised since the 1950s for the prevention and/or arresting of corrosion of steel in concrete caused by chloride ions or by a combination of chloride ions and carbonation. It is well established that the process achieves an immediate reduction in corrosion rate by shifting the steel/concrete potential but that, in addition, the cathodic reactions at the steel/concrete interface increase the alkalinity (raise the pH) by hydroxyl ion (OH^-) generation and drive chloride ions (Cl^-) away from the steel as a result of the negative charge on the ions being repelled by the negative polarity of the reinforcement and attracted to the positive polarity of an installed anode. Thus Cathodic Protection of steel in concrete has always incorporated some degree of Realkalisation (increase in pH) and Chloride Extraction (re-distribution of chlorides). Both these processes reduce the risk of corrosion of steel in the concrete and also achieve further protection.

Cathodic Protection is applied to chloride-contaminated and carbonated concrete after corrosion has commenced and requires 10–20 mA/m^2 current applied to the steel to overcome the pitting corrosion. This current is applied virtually continuously, to provide, typically, a system life of 20–25 years.

Cathodic Prevention is a form of Cathodic Protection applied to new structures and particularly to structures where it is known that they will in future become chloride-contaminated and carbonated. As pitting has not yet commenced the current required is usually 1–2 mA/m^2 of steel. This is applied continuously, typically for a system life of 10–40 years. The cathodic prevention system prevents the onset of corrosion by maintaining a high pH at the steel/concrete interface and by preventing the chlorides (Cl^-) from reaching the interface.

The more recent techniques (1960s and later) of Chloride Extraction and Realkalisation form the subject of this present state of the art report. These use much higher currents, typically 1000–2000 mA/m^2 and are applied temporarily for short periods of generally 5–15 weeks for Chloride Extraction and 5–15 days for Realkalisation. These high current short period processes are intended to achieve permanent modification of the concrete chemistry by virtue of the cathodic (protection) reactions, the movement of ions in a d.c. field and absorption of ions from external electrolytes in contact with exposed surfaces of the concrete.

Cathodic Protection and Cathodic Prevention have a long term track record in steel reinforced concrete structures. State of the art for these techniques has already been laid down in a CEN-Standard (prEN 12696 'Cathodic protection of steel in concrete. Part 1. Atmospherically exposed concrete'). Owing to the lack of effective acceptance criteria and unanswered questions regarding possible side effects and longevity, up to now it was impossible to formulate a standard for Chloride Extraction and Realkalisation. On the other hand, valuable data related to these topics are available from laboratory research as well as from practical cases. Therefore, it is obvious and necessary to compile this information in a technical report. This work has been carried out by a Task Group* of the EFC Working Party 11 on Corrosion of Reinforcement in Concrete and has resulted in the present report.

The document, which has been prepared by J. Mietz with considerable contribution from J. Tritthart, starts with information on mechanisms of steel corrosion in concrete which is necessary for the understanding of electrochemical remediation techniques. Based on this information the principles of Electrochemical Realkalisation and Chloride Extraction are described and experience from laboratory experiments and practical application is added. Important issues regarding side effects are discussed and aspects for the assessment of these techniques in practical rehabilitation are considered.

The report is aimed at closing the gap between laboratory testing and execution of these processes in practice. For this reason the information given in the report is valuable for researchers as well as practitioners. Civil or structural engineers involved in concrete rehabilitation work will benefit from the document in designing electrochemical rehabilitation projects; at the same time the contractor will be informed about the durability of the measures and acceptance criteria.

B. Isecke
Member of the Board of Administrators of the European Federation of Corrosion

---

* The members of the Task Group were as follows:

B. Elsener (Switzerland)

J. Mietz (Germany) – Convenor of the Task Group and chairman of the EFC-WP11.

C. Page (UK)

R. Polder (The Netherlands)

J. Tritthart (Austria)

Ø. Vennesland (Norway).

1
Introduction

Steel or prestressed steel in reinforced concrete structures is protected from corrosion by a passive layer on the steel surface resulting from the high alkalinity of the concrete environment. The long-term durability of this protection against corrosion is related to the conditions necessary for the stability of this passive layer. However, depending on concrete quality, workmanship and constructional characteristics, the passivation effect of the concrete pore solution can deteriorate in certain environmental conditions. In addition to penetration of chloride ions which can locally destroy the passive layer, carbonation of the concrete caused by carbon dioxide from the air can also lead to depassivation and as a result corrosion of the steel may occur.

As corrosion reactions are electrochemical processes, corrosion can be avoided or at least minimised by preventing one of the single sub-processes of the total reaction [1]:

- the anodic process
- the cathodic process
- the electrolytic charge transfer between anode and cathode.

Electrochemical methods stop (or slow down) the anodic and/or cathodic process by polarisation of the steel and by chemical changes in the concrete that result from the polarisation. In order to avoid disadvantages of conventional rehabilitation measures of the patch-repair type, various electrochemical techniques for corrosion protection of reinforcing steel have recently been developed and successfully applied. The main characteristics of the following three most important methods:

- electrochemical realkalisation (RA),
- electrochemical chloride extraction (CE)
- cathodic protection with impressed current (CP)

— all based on cathodic polarisation of the reinforcement — are summarised in Table 1.

Table 1. Characteristics of electrochemical rehabilitation treatments

	RA	CE	CP
Protection objective	Passivating environment	Passivating environment chloride removal	Protection potential
Duration of polarisation	Temporary (3–14 days)	Temporary (6–10 weeks)	Permanent
Current densities*	$0.8 - 2\,A/m^2$	$0.8–2\,A/m^2$	$3–20\,mA/m^2$

*Related to the concrete surface.

2
Scope

This state of the art report describes principles and mechanisms of two electrochemical rehabilitation techniques for reinforced concrete structures: electrochemical realkalisation and electrochemical chloride extraction. Furthermore, possible side effects as well as examples and possibilities for application will be presented. *It does not deal with cathodic protection* as this technique has been already the subject of a CEN standard prepared by TC 262 (prEN 12696 'Cathodic protection of steel in concrete. Part 1. Atmospherically exposed concrete').

3
Mechanisms of Corrosion of Steel in Concrete

3.1. General

During hardening of concrete a solution of alkali hydroxides and alkaline-earth hydroxides is generated within the pores. This solution is characterised by a high pH-value between 12.5 and 14 and provides reinforcing steel with excellent corrosion protection. However, under certain conditions the long-term durability can be lost.

The deterioration process of concrete structures resulting from corrosion of the rebars can be divided into an initiation phase and a propagation phase (Fig. 1) [2]. The end of the initiation phase (time t_d) is given by the depassivation of the steel. The time of corrosion propagation is given by the admissible degree of loss in cross section due to corrosion and by the corrosion rate. Durable reinforced concrete structures can thus be achieved by a very long initiation phase and/or a negligibly low corrosion rate. In the design stage, of new constructions, the engineer will try to achieve a long initiation phase by defining concrete cover and quality, whereas, in the case of rehabilitation a prolongation of service life can be achieved by a reduction in the corrosion rate.

3.2. Carbonation

One mechanism which leads to loss of corrosion protection is carbonation of the concrete. Carbonation is the result of the interaction of carbon dioxide in the atmosphere with alkaline hydroxides in the concrete. The ingress of CO_2 gas from

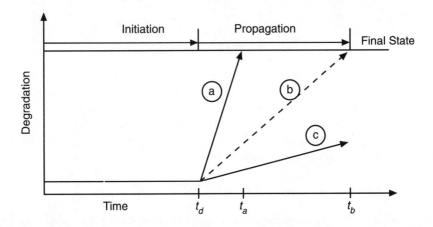

Fig. 1 Development of corrosion of steel in concrete with time [2]. t_d = time of passivation; t_a, t_b = time to reach final state with corrosion rates a and b respectively; a,b,c, = corrosion rates (a>b>c).

the surface depends mainly on the concrete porosity, the relative humidity of the permeable pore structure and the pore structure itself. Carbon dioxide dissolves in water to form an acid and this neutralises the alkalis in the pore water, to form mainly calcium carbonate. As a result, the pH-value decreases to a value below 9 and this will lead to depassivation of the steel surface. Carbonation is a function of cement chemistry, cement content, porosity of concrete, and exposure conditions. These factors will control the pore structure and thus the ease with which CO_2 can penetrate the concrete. Carbonation induced corrosion is promoted by low concrete cover, poor compaction, high water/cement ratio, low cement content, and poor curing conditions.

3.3. Chlorides

The second contaminant that may render corrosion protection ineffective even in an alkaline environment is chloride, which, when present above a certain critical concentration at the steel surface, locally destroys the passive layer. Chlorides can penetrate from sea water or de-icing salts in the concrete environment or be present in the original concrete mix (mixed-in chloride). In the past, calcium chloride (or a mix of calcium and sodium chloride) was used as a set accelerating agent when concrete was prepared at low temperatures or to speed up the production of precast units. Unwashed marine sand or even sea water have also frequently been used. These practices are now forbidden in most countries but corrosion might nevertheless still develop many years after casting, as a result of a combined effect of chlorides and carbonation. It is, however, primarily the subsequent entry of chloride that is of practical importance. Here the most affected reinforced concrete constructions are those close to coastal regions that are directly exposed to sea water or salt-water spray and constructions exposed to de-icing salts. Occasionally, chloride contamination is caused by PVC combustion gases as this produces hydrogen chloride gas which dissolves in water to form hydrochloric acid.

3.4. Concrete Damage

If either the carbonation front or the critical chloride concentration has reached the reinforcing steel protection will be lost and, since normally both oxygen and moisture are available, the steel is likely to corrode. The corrosion reaction is an electrochemical process which can be divided into two sub-processes which may proceed on the steel surface either side by side or locally separated from each other. The actual dissolution of iron takes place at the anode:

$$Fe = Fe^{2+} + 2e^- \tag{1}$$

The free electrons are transported through the steel to the local cathode. In an alkaline or neutral, oxygen-containing electrolyte they are consumed by the oxygen reduction reaction according to:

$$^1/_2\,O_2 + H_2O + 2e^- = 2\,OH^- \tag{2}$$

The total reaction of the single steps leads to the intermediate corrosion product $Fe(OH)_2$ by

$$Fe^{2+} + 2\,OH^- = Fe(OH)_2 \tag{3}$$

which further oxidises to Fe^{3+}-oxides and hydroxides (rust). These reaction products formed by the corrosion process are greater in volume than the steel that has corroded, thus inducing bursting stresses in the concrete around the bar. This in turn causes cracking of the concrete with time. The different forms of damage can vary according to the spacing and size of the rebar and its cover and thickness; from thin longitudinal cracks, delamination with little visible external signs of distress, to spalling off at corners. It should be mentioned that in the presence of high chloride concentrations soluble iron chlorides, green rust or other non-voluminous products that do not lead to spalling can also be formed.

4
Electrochemical Realkalisation

4.1. Principles

The most effective method to stop or prevent carbonation induced reinforcement corrosion and thus avoid costly repair of buildings and structures is to eliminate the causes of corrosion reactions, that means the re-establishment of the passivity that has been lost. For reinforced concrete structures suffering from carbonation, electrochemical realkalisation has been developed as a rehabilitation treatment [3–5]. This method is aimed at stopping reinforcement corrosion with minimum disruption to the structure and minimal risk of further corrosion. The concrete in the vicinity of the reinforcement is re-alkalised and thus a new passive layer is generated providing renewed protection of the steel surface.

The principles of electrochemical realkalisation are shown in Fig. 2. The technique involves passing a current through the concrete to the reinforcement by means of an externally applied anode which is attached to the concrete surface. Sodium carbonate is generally used as the electrolyte covering the concrete surface. In recent investigations it has been shown that lithium hydroxide could also be used as an appropriate electrolyte [6]. To avoid short circuits, cracks and other defects must be sealed before electrochemical realkalisation is applied (see also Section 5).

The electrode outside the concrete (anode) and the reinforcement inside acting as cathode are connected to a direct current (d.c.) source. During the treatment various processes take place as described below.

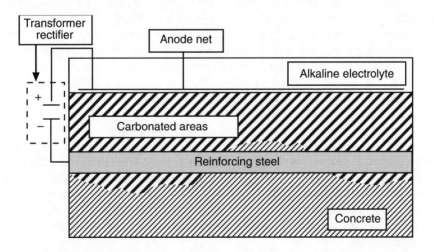

Fig. 2 *Principle of electrochemical realkalisation.*

4.1.1. Electrolysis

As a result of the electrical potential difference between the anode and the cathode, water is reduced to hydroxide ions at the reinforcement. After the available oxygen has been consumed according to (4) (which is likely to take less than 24 h), the cathodic reaction produces hydroxide and hydrogen gas according to (5)

$$^1/_2\,O_2 + H_2O + 2\,e^- = 2\,OH^- \qquad\qquad (4)$$

$$2\,H_2O + 2\,e^- = H_2 + 2\,OH^- \qquad\qquad (5)$$

At the external anode, water or hydroxide is oxidised to oxygen and hydrogen ions according to eqns (6) and (7), respectively.

$$2\,OH^- = {}^1/_2\,O_2 + H_2O + 2\,e^- \qquad\qquad (6)$$

$$2\,H_2O = O_2 + 4\,H^+ + 4\,e^- \qquad\qquad (7)$$

These reactions are referred to as electrolysis, which results in a pH increase at the reinforcement, and this is the most important process for realkalisation. For non-inert metal anodes oxidation has to be considered as the anodic reaction (see Section 7.3.).

4.1.2. Electromigration

Aqueous solutions of salts, bases, acids etc. are electroconductive. These substances are present in dissolved form in a dissociated state, i.e. they have been split into electrically charged particles (ions). Upon application of direct voltage ions begin to move due to the force exerted on these charge carriers by the field strength that has been built up between the electrodes. This process is called electromigration.

The electric current that flows through the solution at a given voltage and temperature is proportional to the number of ions present in the solution and, furthermore, dependent on the velocity at which these are able to move towards the electrodes. The total conductivity of an electrolyte is an addition of the partial conductivites of the individual ions. The contribution of a certain ion to the total current is called the Transference (or Transport) Number. For a given ion this number will be larger the larger the quantity and mobility of the ion.

The interactions that apply to aqueous solutions are, basically, also valid for concrete because in concrete the transport of electric current is facilitated almost exclusively by migration of ions dissolved in the water-filled capillary pores of the concrete. As a result of the applied electrical field between the reinforcement and an external anode, negatively charged ions (e.g. chloride, hydroxide) will move towards the anode and positive ions (e.g. sodium, potassium) will move to the cathode, i.e. the reinforcement.

4.1.3. Absorption

Because of capillary effects dissociated alkaline solution will be absorbed from the concrete surface. The absorption effect strongly depends on the moisture condition of the concrete, i.e. whether it is dry or wet, and on the pore structure of the particular concrete. Results from a demonstration project have shown that the absorption of alkali is not very important when applying current densities and treatment times commonly used in practice [7].

4.1.4. Diffusion

Diffusion processes take place when there are different concentrations of the various compounds. Mathematical estimations indicate that with respect to the timescale of the practical treatment, such concentration gradient driven diffusion can be neglected as an effective contributor to realkalisation [8]. The rate of diffusion however may be enough to be significant with respect to the service life of the structure. Diffusion of alkalis after realkalisation may decrease the local alkalinity at the steel and consequently reduce the corrosion protection.

4.1.5. Electroosmosis

It has been proposed that the passing of current causes an electrolyte liquid flow into the concrete by a mechanism called electroosmosis. Its significance, however, has not been proved by direct experiments and its magnitude cannot be estimated accurately. Macroscopic liquid transport into the concrete by electroosmosis would be favourable for realkalisation, but unfavourable for chloride extraction. No evidence for electroosmosis has been found in laboratory tests or during testing of cores in a realkalisation demonstration project [7,9,10].

4.1.6. Summary of the mechanisms of realkalisation

From theory, the following model may be set up. Electrolysis produces hydroxide proportional to the charge passing, of which electromigration moves a part towards the anode, which eventually leaves the concrete; the remainder remains near the steel. When a sodium salt is used as electrolyte the sodium enters the concrete and moves towards the steel. The equivalent amount of sodium entering must be equal to the amount of hydroxide remaining (electroneutrality). As the transport numbers of hydroxide and sodium are about 0.8 and 0.2 respectively, the net effect around the steel is that 20% of the charge will result in accumulation of sodium hydroxide near the steel [7]. Any ingress of alkaline material due to absorption or electroosmosis, adds to the above electrolysis/electromigration effect.

Figure 3 shows schematically the effect of the different mechanisms mentioned above.

4.1.7.Long-term effect on steel protection

Hydroxide ions generated at the cathode by electrolysis increase the pH-value around the reinforcement thus providing an alkaline environment for the steel (pH 13–14).

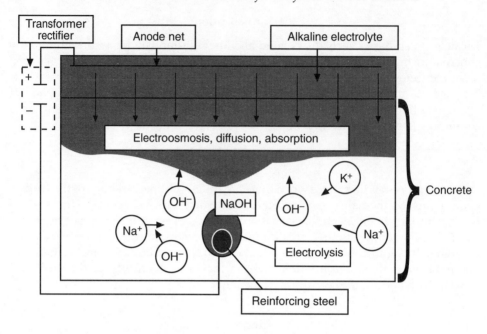

***Fig. 3** Schematic of pH increase due to electrochemical realkalisation.*

This hydroxide may react with carbon dioxide to form sodium carbonate, lowering the pH. The degree of conversion of the total amount of hydroxide to carbonate depends on the rate of penetration of carbon dioxide which, in turn, depends on the permeability of the concrete cover (pore structure, moisture content). It is not clear whether this conversion takes place on a time scale of months or years. The assumption that all hydroxide is converted to carbonate is a 'worst case scenario'. After completion of the conversion to carbonate, further reaction between sodium carbonate and carbon dioxide may occur according to eqn (8):

$$Na_2CO_3 + CO_2 + H_2O = 2\,NaHCO_3 \tag{8}$$

In the state of equilibrium with a constant carbon dioxide concentration of the atmosphere, only small amounts of sodium carbonate will react to form sodium hydrogen carbonate and hence future carbonation leads to only slight decrease of pH [11,12]. In this way, the presence of sodium carbonate in the concrete pores should act as a carbon dioxide trap. Under normal atmospheric conditions, the equilibrium pH is in the range of 10 to 11, probably close to 10.5. However, it may take many years to reach this value. Applying a coating to the concrete after realkalisation will delay this by many years.

4.2. Laboratory Experiments

4.2.1 Mechanisms

In order to seperate and quantify the complex interactions between the physical, chemical, and electrochemical mechanisms and to check the fundamental possibilities and limitations of this kind of electrochemical treatment, laboratory experiments have been carried out within an extended research project using different types of carbonated reinforced mortar specimens [9,10,13]. The so-called mortar electrodes (cylindrical as well as rectangular specimens) were manufactured with water/cement ratios of 0.6 and 0.7 respectively and cement (OPC – ordinary Portland cement) contents between 300 and 470 kg/m^3 to simulate different qualities. Prior to electrochemical realkalisation the test specimens were carbonated under atmospheric conditions.

After polarising with the different current densities and treatment times as commonly used in practical applications, the realkalisation success was at first assessed by means of an alcoholic phenolphthalein solution as colour indicator. Figure 4 shows realkalisation depths developing from the surface vs treatment time with different applied current densities. As expected, the results show increasing realkalisation depths near the surface with longer treatment times. However, significant dependences on the applied current density could not be found. Even realkalisation depths for specimens without electrochemical polarisation were found within the overall scatter range. From these results it can be concluded that the penetration of alkaline electrolyte from outside is mainly controlled by diffusion and absorption. Obviously, the electrochemical polarisation has only a minor effect on the realkalisation progress developing from the concrete surface. On the other hand, a pronounced influence of the applied current density on the thickness of the realkalised layer around the steel bar was observed as is shown in Fig. 5. Because of increasing amounts of hydroxide ions generated with higher current densities, the thickness of the realkalised layer also increases.

The efficiency of the carbon dioxide trap, referred to above, mainly depends on the uptake of sodium ions from the alkaline electrolyte and their subsequent migration towards the reinforcement. Figure 6 gives examples of sodium distribution over the

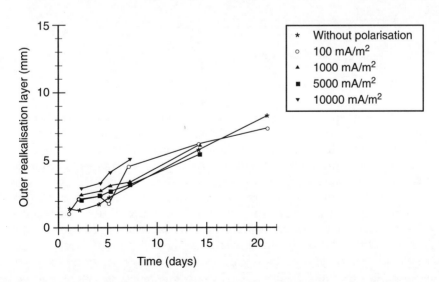

Fig. 4 Realkalisation depths from the surface for different realkalisation conditions [10].

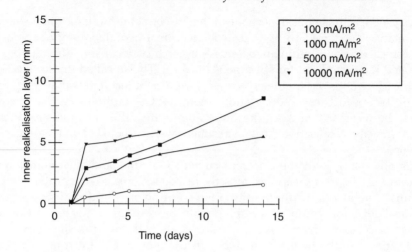

Fig. 5 *Realkalisation depths from the reinforcement for different realkalisation conditions [10].*

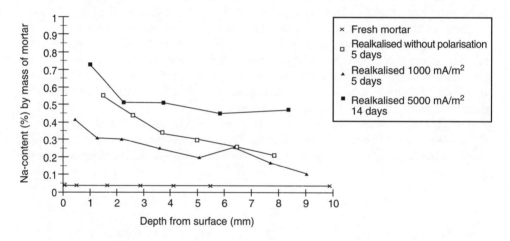

Fig. 6 *Sodium distribution [10].*

radius of different specimens. With depth of cover the sodium content decreases and this agrees well with results reported by Elsener and Böhni [14].

For re-establishment of repassivating conditions an increase of the pH-value adjacent to the reinforcement is necessary. As direct measurements of the pH are difficult to conduct, qualitative alkalinity profiles were determined. Pulverised mortar samples from different depths were mixed with distilled water and the resulting suspensions titrated with HCl using phenolphthalein indicator which procedure excludes any calcium carbonate. The consumption of acid is a measure of the total alkalinity and thus also of the buffer capacity. Figure 7 shows typical results. The circle symbols belong to a partially carbonated specimen. A strong increase from about 4 mm towards the alkalinity level of uncarbonated specimens can be seen. The alkalinity of electrochemically treated electrodes is in the range of those

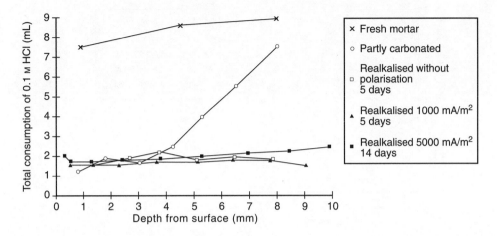

Fig. 7 *Alkalinity profiles [9].*

carbonated independent of the realkalisation conditions. This means that increase of alkalinity with the object of re-establishing the buffer capacity of a sound concrete cannot be achieved by electrochemical realkalisation.

Laboratory investigations using lithium hydroxide as alkaline solution have been reported by Sergi *et al.* [6]. It was found that LiOH was at least comparable to Na_2CO_3 in terms of the pH enhancement near the steel cathode and was thought to offer extra benefits as it provided a high pH zone extending 10–15 mm from the mortar surface while introducing an inhibitor for the alkali-silica reaction into the material.

Figure 8 shows typical lithium concentration profiles immediately after realkalisation. Most showed a significant concentration of lithium throughout the length of the specimen with higher concentrations near the steel and particularly near the surface. The penetration depth of the lithium ions from the surface did not appear to be greatly influenced by either the current density or the time of exposure which is in agreement with the results mentioned above obtained in Na_2CO_3, i.e. that the penetration of alkaline electrolyte from outside is mainly controlled by diffusion and absorption. In most cases measured concentrations of lithium ions were significantly greater than those of sodium and potassium ions.

4.2.2. Assessment of effectiveness

Assessment of the effectiveness of the realkalisation procedure by spraying a sample of the realkalised concrete with phenolphthalein indicator can only indicate whether the pore electrolyte of the sample has exceeded a pH-value of about 9.5, but cannot differentiate pH values. Therefore, a better and more accurate assessment procedure is required. Mietz *et al.* [9] subjected specimens of different treatment conditions to subsequent anodic polarisation tests. These tests were carried out in the galvanostatic as well as the potentiostatic mode. Figure 9 shows as an example current vs time curves of fresh, carbonated and realkalised specimens (the realkalised specimens were treated 3, 5 and 7 days with 500 mA/m²) during potentiostatic polarisation. The tests were carried out at a constant potential of 600 mV with respect to the

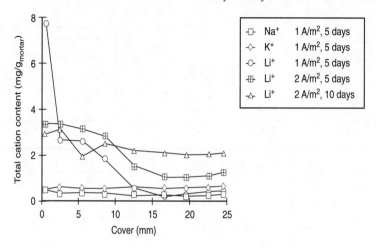

Fig. 8 *Lithium, sodium and potassium concentration profiles after realkalisation in 2M LiOH [6].*

standard hydrogen electrode. This test is a proven method for assessing detrimental effects of concrete admixtures (it has to be mentioned that in corrosion testing of admixtures the potential is 100 mV lower than in the tests described above). On the one hand, the potential is below the oxygen evolution potential and, on the other hand, passive steel will cause no current flow, unless depassivation takes place. Fresh mortar specimens show a passive behaviour with current densities in the range of zero while the reinforcing steel in carbonated samples is depassivated leading to a substantial iron dissolution current density. The specimens with 3 or 5 days of realkalisation cannot be classified as repassivated as there is a more or less pronounced dissolution current density. However, after a 7-days realkalisation treatment the current density decreases to negligible values. Although all comparable specimens had shown red-colouring around the steel in the phenolphthalein test, the

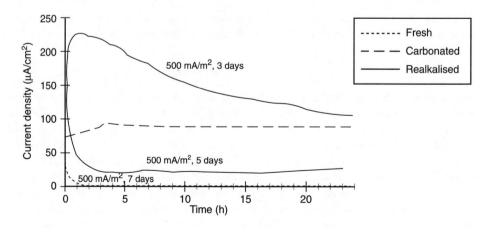

Fig. 9 *Results of potentiostatic anodic polarisation [10].*

polarisation, both potentiostatically and galvanostatically, indicates that especially with short treatment times and low current densities, passing the phenolphthalein test (i.e. red-colouring of the mortar around the reinforcement) is not a sufficient indicator for real repassivation. Sergi *et al.* [6] also deal with the problem of assessing the effectiveness of the realkalisation treatment. These authors developed a two-stage technique to improve the assessment procedure. The first stage involves the application of several pH indicators, covering a range between pH 10 and pH 13, to freshly broken, recently realkalised mortar or concrete. This establishes a pH profile around the steel and within the cover and represents an improvement over the phenolphthalein test which simply determines a band of concrete achieving a pH higher than about 9.5. The second stage involves subjecting the steel reinforcement to galvanostatic anodic polarisation while monitoring its potential in order to determine the degree of repassivation of the steel. Trials on cores including reinforcement bars from actual structures indicated that it would be possible to adopt this technique as a performance criterion during various stages of the realkalisation process.

4.3. Field Experiences

The first approaches to realkalisation were based on the principle of pure diffusion [15], that is, if an external layer of very alkaline mortar is applied on carbonated concrete, hydroxide ions will diffuse into the concrete due to the concentration gradient. However, the practical results were not always as successful as expected, mainly due to the slowness of the diffusion process. Electrochemical realkalisation was commercially introduced in 1987 [4]. The idea was to re-establish the corrosion protective qualities of concrete by increasing the alkalinity of carbonated concrete up to a protective pH. The normal procedure uses steel mesh as the external anode and 0.5 to 1.0M sodium carbonate solution as electrolyte. Although there are a number of reference structures which have been treated with this technique only a few reports of practical experiences exist [12].

One of the first realkalisation projects was the State Bank of Norway in Stavanger [12]. A survey prior to rehabilitation revealed that approximately 70% of the vertical reinforcement was corroding due to carbonation. The project was carried out in 1988 using 1.0 M sodium carbonate solution. Data on duration and current density of the treatment are not available. Immediately after the treatment, the realkalisation depth was measured using phenolphthalein indicator, and potential measurements with cast-in reference electrodes were made. After 2, 4 and 5 years further surveys were carried out including potential measurements, indicator tests on drilled cores and sodium analysis of cores. The results obtained after 5 years have shown that apart from areas vulnerable to washing-out of the electrolyte, there is no sign of significant reduction of the alkali content in the realkalised concrete.

Phenolphthalein tests and sodium analyses of drilled cores 2 and 4 years after a realkalisation treatment of a church in Switzerland carried out in 1990 have shown that the alkali content adjacent to the reinforcement is sufficiently high to establish a protective pH [12,16] whereas in the concrete between the reinforcement, the sodium content is significantly lower. The sodium analyses in this example revealed one problem of this kind of assessment criterion. Thus, some sodium values indicate an apparent increase in the sodium content from 2 to 4 years. This is, of course, not a real increase, and is explained by the 4-year cores being drilled adjacent to reinforcement bars, while the 2-year cores were drilled between bars.

In a demonstration project in the Netherlands it was found that the realkalisation treatment produced a pH above 14 close to the steel immediately after the treatment [7]. However, as a result of diffusion of hydroxide the pH decreased to lower values in 3 to 4 months. In some cases the effect of the treatment was assessed by determining the pH value in concrete pore water samples. Figure 10 shows the distribution of pH values before and after the realkalisation. In this case pH-values above 12 guaranteed adequate corrosion protection of the reinforcement after the electrochemical treatment [17].

It is surprising that few documented site jobs on electrochemical realkalisation can be found in the literature despite the quite frequent use of this technique in practice. Elsener *et al.* were the first to present results of a well characterised and documented realkalisation job, a reinforced concrete building facade [14,18,19]. The electrolyte used was 1.0M sodium carbonate solution. A current density of *ca.* 2 A/m² was imposed for 12 days. Prior to and after the realkalisation treatment different types of condition assessment techniques were employed, i.e. potential mapping, galvanostatic pulse measurements, resistivity, porosity and pore solution composition. Based on these results the two main practical problems with realkalisation, the control of the efficiency (repassivation of the steel) and the long-term durability of the treatment were discussed.

The high cathodic current during electrochemical realkalisation shifts the potential of the steel in concrete to very negative values [14,18,19]: immediately after switching off the current, values of <–1.1 V(SCE) were measured, after 1 day the potentials were in the range of –0.7 to –0.9 V(SCE). The potential over the treated area was quite homogeneous already after 7 days and no signs of the 'hot spots' present before the treatment could be observed. The depolarisation measurements showed that

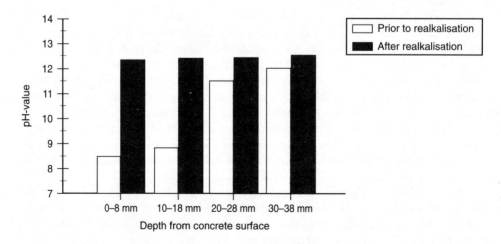

Fig. 10 *Distribution of pH-values before and after electrochemical realkalisation [17].*

after about one month the potentials had stabilised around –0.2 V(SCE), a value that corresponds to passive steel in wet, alkaline concrete. The authors mention that, in contrast to electrochemical chloride removal, measurements of the steel potential from the concrete surface — even after complete depolarisation — cannot be interpreted easily in terms of repassivation since at the steel the pH and the corrosion state are changing and the concrete cover changes in moisture and resistivity. Potential values around –0.2 V(SCE) and the presence of a homogeneous potential field after the treatment are strong indications of repassivation. On the other hand; analysing the sodium profiles gives no direct indication on the steel repassivation or of the pH value around the rebars.

4.4. Long-term Behaviour

As described above (Section 4.3.) investigations on realkalised structures have shown that there is no sign of significant reduction of the alkali content in realkalised concrete. Nevertheless, with time it is likely that there may be a gradual reduction in the highly alkaline regions around the bars to achieve equilibrium with areas which were not realkalised.

Mattila *et al.* [20] examined the leaching of alkalis by means of sodium concentration measurements during an accelerated weathering test. On the basis of these measurements it was stated that alkali concentration may be lowered by intense weather exposure (cyclic wetting and drying). The effect of leaching will affect primarily the zone near the surface. From their results it can be concluded that corrosion protection in realkalised concrete may be endangered by leaching only in low cover areas (cover < 5–10 mm) in structures where weather exposure is intense.

It has also to be mentioned that in spite of the re-establishment of a high hydroxide level around the reinforcement the recovery of the alkaline buffer capacity which exists in non-carbonated concrete (based on $Ca(OH)_2$) will not be attained. Laboratory tests on the long-term durability of realkalised specimens have confirmed this fact [9,10]. Although passive behaviour was detected immediately after the realkalisation treatment, several specimens indicated depassivation after several months of exposure. Obviously, the repassivation effect was rather weak with the consequence that the passive layer around the reinforcing steel could be destroyed.

The durability of the electrochemical realkalisation depends on the possibility of maintaining a passive layer on the steel surface, i.e. neutralisation of the concrete pore solution must be avoided. This could be achieved by the equilibrium pH of the $Na_2CO_3/NaHCO_3$ solution which will be established under atmospheric exposure conditions (see 4.1.). However, the durability of the corrosion protection in the presence of chlorides has to be studied further [7,18,21].

Results from a test site in Switzerland [19] have shown that half-cell potential mapping is a possibility for monitoring the long-term behaviour after the treatment. In the case described the measured potentials remained nearly constant and homogeneous after more than one year.

As there is concern about adverse effects of the treatment on other properties of the concrete (possible reductions in bond strength, possible increased risk of alkali-aggregate reaction resulting from higher sodium concentrations in the concrete, and changes in the microstructure of the concrete which may modify mechanical and durability properties) different experimental investigations have been made [22–27]. These various reports and papers address results of both realkalisation as well as chloride extraction. Conclusions will be discussed in Section 6 (Side Effects, pages 35–40).

5
Electrochemical Chloride Extraction

Chloride-contaminated concrete is usually repaired by removing the outer chloride-containing concrete zone all the way to behind the reinforcement and by replacing the old concrete. This approach has several disadvantages. For example, possible problems result from the partial replacement of the concrete, such as serious changes in structural analysis or damage to the reinforcement when the old concrete is removed. Also, the impact on workers and the environment (dust, noise, etc.) is considerable. If the repair removes only damaged areas but not all chloride-contaminated concrete then this measure has only temporary effects. In a lot of cases the non-destructive electrochemical chloride extraction (CE) — sometimes also called electrochemical chloride removal or desalination — offers an attractive alternative compared to removing all chloride-contaminated concrete from the structure. In case of heavily contaminated columns, foundations or structural members, for which a monolithic behaviour is essential, it is only with electrochemical methods (CE or CP), that durable protection is possible at all.

5.1. Principles

Electrochemical chloride extraction is based on migration of ions. As with electrochemical realkalisation the reinforcement is connected to the negative pole and the anode — placed within an electrolyte covering the concrete surface — is connected to the positive pole of a d.c. source (Fig. 11). Within the electrical current field negatively charged chloride ions move along the current flow lines from the reinforcement representing the cathode towards the anode outside the concrete. The movement of chlorides in concrete is mainly caused by migration, since the transport

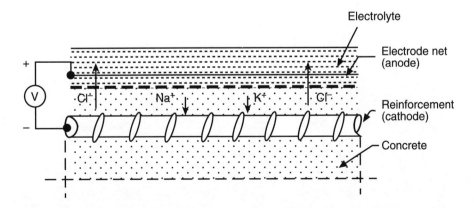

Fig. 11 *Principle of electrochemical chloride extraction.*

caused by diffusion and convection is so low at the relatively high current densities that are applied for electrochemical chloride removal, that it can be neglected [28].

The rate at which chloride is removed is directly proportional to the current flow through the concrete. For a given voltage this is highest when the electrical resistance of the concrete is very low, i.e. when the concrete is water saturated. The part of current which is transported by chloride ions is proportional to the chloride concentration [29].

As chloride is transported along the lines of current flow that have formed in the concrete, the spatial current distribution between reinforcement and the anode mesh placed on the concrete surface is of importance. Figure 12 shows a schematic representation of the streamline pattern in an inhomogeneous field with equal resistivity everywhere between a bar and a surface, i.e. in a configuration similar to that in electrochemical chloride removal [29].

As can be seen from Fig. 12, the streamlines are shortest between the vertical line connecting the bar (cathode) and the surface (anode) and longest starting from the back of the bar. The shorter the streamlines the stronger is the force acting on the ions, and thus the greater the migration speed of the ions (local current strength). For chloride removal from reinforced concrete this means that the current strength, and hence the migration speed of the ions, is highest in the zone directly above the reinforcement while it becomes less as the lateral distance from the reinforcement increases. This means that within the cover chloride is extracted most slowly from the mid-area between two bars, and comparatively little chloride will be removed from behind the reinforcement.

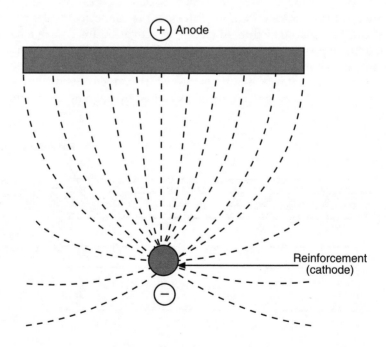

Fig. 12 Streamlines in an inhomogeneous field between bar and surface [29].

From the above interactions it becomes quite obvious that differences in the thickness of the cover also play an important role. Within the area to be repaired the highest current densities must be expected in zones where the cover is thinnest (the electrical resistance between reinforcement and anode is lowest there). This means that the efficiency of chloride removal in concrete of varying cover thickness will be relatively low in areas with a thick cover and relatively high in areas with a thin cover. Assuming that an area existed without cover, that is to say the reinforcement ran on the concrete surface, a short circuit would result and the entire current would flow here. Thus, special care is needed for all areas of low resistance, e.g. where there is a danger that the outer electrolyte might get directly to the steel surface. Therefore, cracks and other defects must be sealed before electrochemical chloride extraction is applied.

The electrical potential difference between the anode and the cathode causes the same reactions (eqns (6) and (7)) as for electrochemical realkalisation:

$$2\ OH^- = {}^1\!/_2\ O_2 + H_2O + 2\ e^- \tag{6}$$

$$2\ H_2O = O_2 + 4\ H^+ + 4\ e^- \tag{7}$$

$$2\ Cl^- = Cl_2 + 2\ e^- \tag{9}$$

At the same time around the reinforcement hydroxide ions are generated by electrolysis.

$${}^1\!/_2\ O_2 + H_2O + 2\ e^- = 2\ OH^- \tag{4}$$

$$2\ H_2O + 2\ e^- = H_2 + 2\ OH^- \tag{5}$$

The increase in OH^- and the simultaneous reduction of the chloride content near the rebars enables repassivation of the steel surface if the chloride content can be reduced below the critical threshold value.

Different solutions are used as the outer electrolyte. The most widely used are saturated calcium hydroxide, sodium borate, sodium hydroxide or simply tap water. When an alkaline solution is used, the OH^- ions present at the anode are converted into oxygen gas and water molecules, according to eqn (6). As a result, the pH-value of the electrolyte decreases. When an approximately neutral unbuffered electrolyte such as tap water is used, the water is decomposed at the anode and oxygen-gas and hydrogen-ions are formed, according to eqn (7). Provided that the outer electrolyte does not contain any salts, this is the only reaction that occurs until anions have moved in from the concrete and are discharged. The pH-value shifts to the acidic range because the H^+-ions that have formed migrate in the direct voltage field towards the negative pole and meet OH^- and Cl^- ions moving in the opposite direction [30]. The OH^- ions are neutralised to water, and hydrochloric acid is formed with the Cl^- ions. According to eqn (9), the Cl^- ions arriving at the anode are discharged and form chlorine gas.

Both the acidification of the electrolyte and the formation of chlorine gas are often regarded as undesirable. An electrolyte that has turned acidic may attack the concrete, and chlorine gas is very toxic and thus hazardous to health. The use of an alkaline electrolyte, such as a saturated $Ca(OH)_2$ solution or a sodium borate solution, etc., can prevent the acid attack on the concrete and the development of chlorine gas. At pH-values above 7 practically no chlorine gas is formed because the reaction according to eqn (6) takes place more easily.

5.2. Practical Experiences

The first trials of electrochemical chloride extraction were conducted in the USA from 1973 to 1975 by Lankard *et al.* (Battelle Columbus Laboratories, Columbus, Ohio) [31,32] and by Morrison *et al.* (Kansas Department of Transportation, Topeka, Kansas) [33].

Both studies focused on the question of whether the method was suited in principle for chloride removal. Quick chloride removal was regarded as essential to keep the necessary closure of the highway lanes and the concomitant obstruction of traffic as short as possible. Lankard *et al.* determined in laboratory tests carried out on concrete cylinders and test slabs that had been prepared with the addition of chloride that 100 V were required as at 50 V chloride removal did not happen fast enough. Platinum-coated titanium wire was reported to be the optimal anode material. An ion exchanger, which had been impregnated with a saturated $Ca(OH)_2$ solution, was placed between the concrete surface and the anode. The ion exchanger was employed to avoid the formation of chlorine gas, i.e. the Cl^- ions that escaped from the concrete into the electrolyte were exchanged for OH^- ions.

Morrison *et al.* operated at voltages of up to 220 V. The formation of chlorine gas was prevented by these authors by using copper wire as anode material, so that the copper was dissolved as copper chloride. The dependence of the efficiency of chloride withdrawal on the distribution of streamlines in the concrete was described in detail. The reduction of the chloride content was found to be less behind the reinforcement and in the middle between two steel bars than directly above the reinforcement. The efficiency increased within the cover from the outside to the inside and was highest in the zone immediately above the reinforcement.

The authors of both publications draw the conclusion that, basically, the method is effective, but that it also has a number of drawbacks. For example, the energy efficiency factor was rather poor, and according to calculations by Lankard *et al.*, not more than about 6% of the charge flowing through the concrete was transported by chloride. Morrison *et al.* reported that, as a result of the discharge of alkali ions, the reinforcement had been coated with alkali metal. The subsequent anodic dissolution of the alkali metal at the anode after the power cut-off caused a temporary potential difference in the concrete which was feared would encourage the Cl-ions that had remained in the concrete to flow back to the reinforcement. The temperature of the concrete rose from 24°C to about 52°C within 24 h. Furthermore, a significant increase in concrete porosity was described, as a result of which a five-fold increase in permeability was noted [33]. After these two studies the method was not further investigated for some time. This was probably due to a number of factors, such as the safety risk associated with the high direct voltages required, the low energy efficiency factor, as well as potential damage (loss of the bond between steel and concrete, cracks caused by gas pressure resulting from the formation of hydrogen at the reinforcement, and/or temperature rise, higher concrete porosity, etc.)

In Europe, a patent entitled "Removal of Chlorides from Concrete" was published in 1986 [34], but not the studies on which it had been based [35]. The patent specifications include the use of a material that can take up the outer electrolyte, adheres well to non-horizontal and vertical surfaces, and can be easily removed. The material most commonly used is paper pulp. It is applied by gun spraying, and the anode is then embedded therein. The voltage applied between anode and cathode of preferably ≤30 V is considerably lower than in the tests described above, however, the procedure lasts much longer and its duration is determined by the progress in chloride removal. This variation of electrochemical chloride removal and the method of electrochemical realkalisation [36] is called the NORCURE™ method. As the patents were exploited for a long time by the "Norwegian Concrete Technology" (NCT); chloride extraction and realkalisation are often simply referred to as the NCT method.

In the USA, the method was studied in greater detail within the "Strategic Highway Research Program" (SHRP) between 1988 and 1993 [37–40]. The studies included a great number of issues of high practical relevance, such as the efficiency of the method, the diffusion of non-removable chloride back to the reinforcement, the bonding strength between steel and concrete, the formation of cracks in the concrete, alkali-silica reaction, formation of hydrogen at the reinforcement, etc. The findings will be discussed in greater detail when these points are dealt with more specifically (see Sections 5.3., 5.4. and 6).

SHRP recommends the use of an activated titanium grid as anode. The electrolyte should be alkaline and have a sufficient buffer capacity to prevent the formation of chlorine gas and damage of the concrete surface. The latter may occur when the pH value of the outer electrode drops to the acidic range. Calcium hydroxide is the electrolyte commonly used in the desalination process. However, this electrolyte proved unable to buffer the pH for any substantial period thus necessitating frequent replacement of the buffer [41,42]. Lithium borate solution has also been used as research had reported that it has the property of counteracting the effects of alkali-aggregate reaction by forming a non-expansive gel. The lithium solution is better than calcium hydroxide at maintaining a high pH [37,41].

The current density recommended by SHRP is between 1 and 5 A/m^2, however, a d.c. voltage of 50 V should not be exceeded for safety reasons. Within this range, 10–30% of the total current can be used for chloride transport. After a total current passage of 600–1500 Ah/m^2, chloride removal becomes inefficient; this point is normally reached after 10–50 days of application. Higher charge quantities and longer periods of application are recommended only if the chloride content in the concrete is extremely high. About 20–50% of the chloride contained in the concrete could be removed by this method.

The SHRP-implementation guide suggests the use of this method when corrosion is in the early stages before the concrete has suffered any major damage. In such cases, the life of a construction can be prolonged by 5–10 years. Altogether, the method is being regarded as useful, and adverse effects (higher concrete porosity, formation of cracks, etc.) are not reported as long as certain recommendations for the implementation of the method are observed (current density ($\leq 5 A/m^2$, max. 1500 Ah/m^2), and if the concrete does not contain any alkali-sensitive aggregates.

Since the early 1990s, more frequent reports on the method have appeared [24,30,37–40,43–58]. In 1990 a conference on "Electrochemical Protection Methods for Reinforced Concrete Constructions" took place at the Swiss Federal Institute of Technology, Zurich, ETH, and electrochemical chloride removal was given much attention [29,44–46]. The conference proceedings include contributions on the basic principles of the process and report cases from Canada and Switzerland where the NORCURE™ method had been applied and the effectiveness of chloride removal could be proven.

The first well documented field application of chloride removal in Europe has been described in a report of the Swiss Federal Highway Agency [59,60]. The results of a two-year field study have shown that about 50% of the initial chloride content was removed within eight weeks (the total charge was 5×10^6 C/m^2). In the treated zones of the structure the half-cell potentials became more positive by about 80–100 mV. It was shown that CE is a feasible technique for heavily chloride-contaminated structures, although very inhomogeneous chloride contamination may require two treatments.

An interesting variation was reported by Rose [56] in which instead of the usual titanium grid, an aluminium foil was used as anode. This aluminium foil should act as an automatic control of the current density. This is needed if the electric resistance in the area to be repaired varies very strongly (places with a very thin cover, faults like cracks or poor compaction) because the chloride is removed very quickly from zones of low resistance (high current density) and very slowly from zones of high resistance (low current density). The self-regulating mechanism of this method is due to the fact that the aluminium foil acts as a sacrificial anode and is dissolved at a speed which is proportional to the intensity of the current flow. This means that the foil will be dissolved sooner and develop holes at places where the current density is high due to the low local resistance of the concrete, than at places of high concrete resistance. The disappearance of the anode immediately above a place of low concrete resistance results in the current density, and thus the effciency of chloride removal, being decreased so substantially that the self-regulating effect will take place.

Within the COST 509 program of the EU "Corrosion and Protection of Metals in Contact with Concrete", which began in 1992 and was completed in 1996, electrochemical chloride extraction, *inter alia*, was investigated by co-operating research institutions in Austria, UK, the Netherlands and Norway [61].

5.3. Chloride Distribution and Efficiency of Chloride Extraction

5.3.1. Chloride distribution before and after the treatment

The most important indicator of how successful chloride removal has been is the extent to which the chloride content has decreased in the concrete. Tests which determine the chloride profile before and after the application of the method are particularly useful because they make it possible to assess whether and where the chloride retained in the concrete is present in dangerously high concentrations, i.e. whether there is a risk of new corrosion by the diffusion of chloride back to the reinforcement. The following examples describe typical results found in practice.

Example 1

Figure 13 illustrates the results obtained by Bennett and Schue during electrochemical chloride removal from the pylon of a bridge [40]. It shows the distribution of chloride after various durations of current flow. It could be calculated from the measuring points in the diagram that within 8 weeks approximately 1.25 kg Cl^-/m^3 were removed at 5 mm depth, ~5.4 kg Cl^-/m^3 at 25 mm depth, and ~3.4 kg Cl^-/m^3 at 55 mm depth. Related to the respective initial chloride contents, this corresponds to ~50%, ~57%, and ~69% at these depths respectively. The differences between the curves demonstrate that chloride removal was most efficient at the beginning and decreased as the period of application grew longer.

Example 2

Figure 14 shows the values before and after chloride removal obtained by Elsener *et al.* from a subway wall in 2 different areas [50]. The dashed lines with the circles

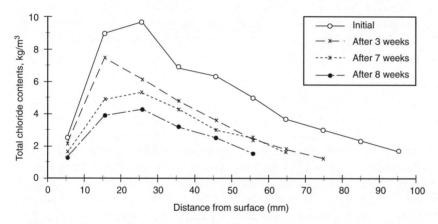

Fig. 13 *Changes in chloride profile during electrochemical chloride removal [46]; (total charge: 610 Ah/m², current density was not constant as the voltage was varied during the treatment, thickness of concrete cover: approx. 75 mm).*

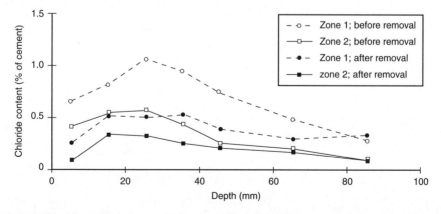

Fig. 14 *Chloride profile before and after electrochemical chloride removal [50] (total charge: approx. 1400 Ah/m², current density changed from 0.75 to 0.3 A/m², thickness of concrete cover: approx. 25–35 mm).*

indicate that a reduction of chloride content by approximately 0.39% (~58% relative to the original content) had been achieved at 5 mm depth, by ~0.53% (*ca.* 50% relative) at 25 mm depth, and by ~0.18% (*ca.* 39% relative) at 65 mm depth. The changes in Cl⁻ distribution were qualitatively the same at the various concrete depths as in Example 1, however, the relative decrease was biggest at the periphery near the surface, and, unlike Example 1, it became less at increasing depth.

Example 3
Figure 15 gives the results obtained by Tritthart from the wall of a reinforced concrete hall with extremely severe chloride contamination [58]. The concrete contained so much chloride because the hall had been used for more than a decade to store loose deicing salts. As can be seen, chloride was withdrawn only down to a depth of about 75 mm. The efficiency of chloride removal differed so widely within this zone that the initial concentration gradient in the concrete was reversed so that already by 40 days the chloride content had increased from the outside to the inside, whereas before Cl⁻ removal it was the opposite. In total, in the 0–10 mm zone 11.5% Cl⁻ (~ 97% of the initial chloride content) were removed, in the 30–40 mm zone 7.3% Cl⁻ (~84%), and in the 60–70 mm zone 2.5% Cl⁻ (~35%).

Example 4
Figure 16 contains results by Polder, measured during chloride removal tests on reinforced concrete blocks of $500 \times 100 \times 100$ mm, which had been stored for 16 years in the North Sea [30]. The blocks contained three smooth reinforcement bars of 8 mm in diameter which ran parallel to the longitudinal axis at different depths (15, 30 and 46 mm concrete cover). The values presented were measured between the surface and the centre of a part of the block where there was no other bar above the mid-reinforcement. As can be seen, the initial Cl⁻ distribution, with values decreasing from the outside to the inside, was qualitatively the same as in Example 3. Nevertheless, Cl⁻ removal was least efficient in the outermost zone and improved as the depth increased (absolute and relative), contrary to the case in Example 3.

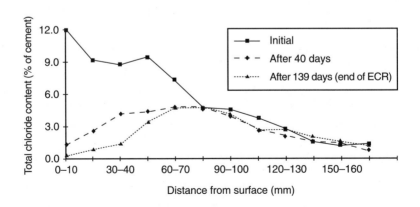

Fig. 15 *Changes in chloride profile during electrochemical chloride removal [58]; (current density: 1 A/m², thickness of concrete cover: approx. 40–60 mm).*

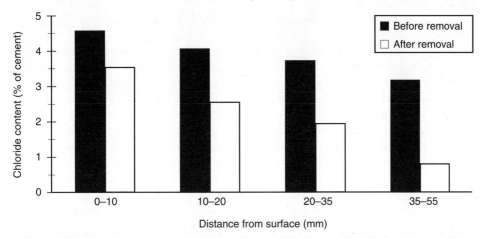

Fig. 16 *Chloride profile before and after electrochemical chloride removal [30]; (total charge: 936 Ah/m², current density: 1 A/m², thickness of concrete of cover: 46 mm).*

Example 5

Figure 17 depicts the changes in chloride content, measured by Tritthart *et al.* from four zones of hardened cement paste specimens as a function of the duration of current flow [51]. The samples had been prepared with the addition of 1% chloride (as NaCl) related to the cement mass. Chloride removal was conducted by placing the cylindrical cement paste samples (length 5 cm; diameter 5 cm) between two chambers which contained the electrodes. After the chambers were filled with the electrolyte (0.5M NaOH), a direct voltage was applied, with the current density kept constant automatically by a voltage regulator. As in Example 4 and contrary to Example 3, chloride removal was most effficient in zone 4, which was closest to the cathode, and least efficient in the opposite sample zone.

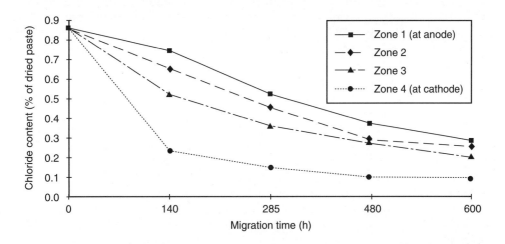

Fig. 17 *Changes in chloride content in different sample sections of cement paste specimens during migration [51]; (current density: 20 A/m²).*

5.3.2. Discussion of examples in 5.3.1.

The examples described above demonstrate that chloride reduction differed considerably from case to case. Initially it was anticipated that the efficiency of chloride removal would diminish within the cover from the inside to the outside. This assumption can be explained by the direction in which the chloride ions move; the chloride ions that migrate from the concrete zone close to the surface into the outer electrolyte are first replenished by other Cl⁻ ions from greater depth, but not at the reinforcement, so that the chloride content ought to decrease more slowly in the peripheral zone than immediately above the reinforcement. This behaviour was actually noted in the Examples 1, 4 and 5 (4 and 5 had been conducted in the laboratory) but not in the practical Examples 2 and 3.

In the case of the laboratory tests, this result can be explained by the test conditions, which were very different from those in practice. Chloride removal from the blocks of Example 4 was conducted by immersing the sample in a vessel which contained the electrolyte (saturated $Ca(OH)_2$ solution) and the anode mesh. The anode enclosed the entire block, so that the chloride could not move away from the reinforcement in one direction only — as is the case in practice — but in all directions. The current density, and thus the efficiency of chloride removal, decreased much more strongly with growing distance from the reinforcement than is the case in practice, where the chloride can move away from the reinforcement in one direction only.

Why the behaviour in Example 5 was identical that in Example 4, although the direction of current flow in the sample was very similar to that in practice (in one direction only), can be explained by changes in the composition of the pore solutions of the samples, which was different from 'normal' concrete. Both electrodes were outside the sample in the electrode chambers which were filled with 0.5M NaOH — an electrolyte which is never used in practice. As the liquid volume of each electrode chamber was almost twice the volume of the entire cement sample, the changes in OH⁻ concentration in the electrode chambers were not big enough to effect any major changes in the OH⁻ concentration of the cement paste sample. This contradicts the situation in practice, where a strong increase in OH⁻ concentration occurs at the reinforcement during CE. Thus, the efficiency of Cl⁻ removal was not changed substantially, so that the decrease in chloride content was slowest in the sample zone at the far end of the migration path. The finding that the relative decrease in the chloride content was less marked in the 5 mm zone than at greater depth in Example 1, compared to the practical Examples 2 and 3, only appears to be a contradiction and is to be attributed to the Cl⁻ distribution before chloride removal. At the beginning of chloride removal the chloride content at 5 mm depth was only about one quarter of the value at 25 mm depth, from where twice as much chloride was extracted than had been initially present at 5 mm depth. All this chloride had to be transported to the outside through the 5 mm zone before the chloride content was able to fall in this zone. The very fact that a decrease was achieved at all at 5 mm depth, although the residual chloride at 25 mm depth still exceeded the initial content in the 5 mm zone, is most probably due to a drastic increase in the efficiency of chloride removal in the outermost zone.

The fact that the decrease in Cl⁻ content was relatively strongest in the deepest zones discussed (55 mm depth) in Example 1, and relatively weakest in Example 2

(65 mm depth) can be explained by the different positions of these zones in relation to the reinforcement. In Example 1 the reinforcement was situated at about 75 mm depth, whereas in Example 2 the cover was only approx. 25 to 35 mm thick; this means that in the latter example the 65 mm zone was located behind the reinforcement, from where chloride can be extracted very slowly only, if at all, while the 55 mm zone in Example 1 was well in front of the reinforcement.

Altogether, these examples demonstrate that the changes in chloride content that occur during chloride removal do not follow a uniform trend and that caution is advised when transferring results from laboratory tests to actual practice. In order to understand the changes observed in an individual case, it is necessary to take into account all the parameters involved, such as Cl^- distribution before and after chloride removal, thickness of concrete cover, the composition of the outer electrolyte, the charge quantity passed through the concrete, etc. However, these examples show that considerable quantities of chloride can be extracted from the concrete cover within a relatively short time although Fig. 15, in particular, makes clear, as already described in Section 5.1. that chloride removal from areas behind the reinforcement is possible only in close vicinity to the reinforcement but not from more distant areas.

5.3.3. Efficiency

The efficiency of CE is the amount of chloride that can be extracted by a certain charge quantity passed through concrete. Polder described it as follows [49]:

$$C_{ex} = I \times t \times CRE \times G/F \tag{10}$$

C_{ex} = extracted amount of Cl^- per m^2 concrete surface

I = current density

t = time of current flow

CRE = Chloride Removal Efficiency Factor (= Cl^- transference number t_{Cl})

G = geometric factor (steel/concrete surface)

F = Faraday's constant.

Equation (10) says that the chloride quantity removed by a certain charge quantity ($I \times t$) is directly proportional to the Chloride Transference Number and a geometric factor.

The geometric factor accounts for the ratio between the concrete surface area (from which the chloride is extracted; see C_{ex} per m^2 concrete) and the steel surface area, i.e. the electrode surface area that passes the current into the concrete.

The concept of Chloride Transference Number was applied for calculation of the efficiency already by Bennett [62]. The Chloride Transference Number is related to the fundamental chemical parameters of ion mobility, charge of the ions in the pore solution and concentration and defines the amount of current carried by the chloride ions in relation to the total current. The Transference Number and thus the efficiency is the bigger the more chloride ions contained in the solution and the lower the concentration of other negatively charged ions. The Chloride Transference Number, and thus the amount of chloride that can be removed by a certain total charge quantity, will be the bigger the fewer OH^- and the more Cl^- ions present in the pore solution, i.e. the higher the chloride content of the concrete.

The efficiency can be calculated from the decrease in chloride content, or simply by the increase in chloride ions in the outer electrolyte, provided the development of chlorine gas is prevented by using a liquid alkaline electrolyte. It is a useful measure of the average efficiency in the specific case of application. However, it is not possible to give a certain range of efficiency of chloride removal applicable to all cases as it will differ from case to case and depend, *inter alia*, on the initial chloride content. This is clearly shown by comparing the examples given in Section 5.3.1. The values given by Bennett *et al.* [37] of an efficiency between 0.1 and 0.3 are nevertheless helpful as a guideline; they tell us that under 'average' conditions between 10 and 30% of the current flowing through the concrete is transported by Cl^- ions. This relates to a maximum flow of current of 1500 Ah/m². The efficiency reduces with increasing amount of charge passed.

However, efficiency does not provide information on the local decrease in chloride content in a particular concrete zone, which — as described in Section 5.3.1. — depends to a large extent on the initial conditions (chloride distribution, situation of reinforcement, etc.). In addition, the Chloride Transference Number, and thus the efficiency of chloride removal, varies between the concrete surface and the reinforcement as a result of changes in the composition of the pore solution brought about by electrode reactions. How big these changes actually are will normally remain unknown as they can be determined only by pore solution tests; but the analysis of pore solution from concrete is not a routine test and has so far been performed only in connection with research projects. For that reason it is not possible in most practical cases to differentiate between bound and free chloride, although only the latter is capable of transporting current. The chloride binding is therefore a very important parameter in CE. In the absence of current it depends, at a given total chloride content, on several factors, such as the type of cement, the w/c ratio, etc. [63,64]. The efficiency of chloride removal is also greatly dependent on the rate at which bound chloride is dissolved. Elsener *et al.*, for example, determined that, when chloride removal had become less effcient over time, it was possible to continue it at increased efficiency after an interruption of current flow for a longer period [50]. The authors assumed that the reason for this was that dissolved chloride had already been removed and that the bound chloride dissolved only very slowly so that the break, during which more chloride was able to dissolve, resulted in an increase in efficiency.

In order to be able to answer questions of this kind, profound knowledge of the mechanisms involved in chloride removal is needed; this, in turn, requires close examination of the changes occurring in the liquid phase.

Chloride removal efficiency also depends to a certain extent on the electrolyte used. Chloride extraction with sodium carbonate electrolyte was found to have a very low chloride removal efficiency [49,54], probably due to the reduced chloride transport number resulting from the increase of sodium and hydroxide concentrations. This was confirmed by pore water analysis before and after CE. Tap water and saturated calcium hydroxide, on the other hand, were found to produce similar removal efficiencies [30,65].

5.3.4. Assessment of the efficiency

At the end of the electrochemical chloride removal the efficiency could be checked by direct measurements (half-cell potential mapping) or by indirect means (charge flow, chloride content) [19].

The residual chloride content in the concrete can be determined from cores taken after the treatment. On structures with inhomogeneous chloride contamination it is difficult to obtain accurate information on the amount of chloride removed. A statistical comparison requiring a number of chloride analyses gives some information about the average reduction of the chloride content but no indication where 'hot spots' with high chloride contents remain on the structure [29,60].

As has been shown in field experiments, the only way to get precise information on the efficiency of the chloride removal is to measure the potential field before and after the treatment (after sufficient time to allow the steels to depolarise). This technique directly indicates if the rebars have been repassivated or if corroding zones still remain [29,60].

5.4. Long-term Behaviour

As chloride is extracted via electromigration primarily from the cover, and even here complete removal is not possible, there is a risk that chloride left in concrete may diffuse back to the reinforcement and encourage further corrosion. How much chloride diffuses and in what time and over what distance depends, of course, on a number of parameters, such as volume and diameter of the capillary pores of the concrete, the extent to which the pores have dried up, chloride binding in the cement, etc. Diffusion happens the more slowly the narrower and the drier the pores. Under identical semi-dry storage conditions, a more porous concrete (high w/c-ratio) dries out more strongly than a concrete with low porosity (low w/c-ratio) so that the changes in diffusion speed are not affected equally in different concretes. Therefore, it is not possible to predict, based on concrete formulation, how fast dissolved ions will be transported by diffusion in air-dried concrete. Bennett *et al.* stored concrete test slabs outdoors for 40 months and observed practically no changes in chloride distribution in slabs where electrochemical chloride removal had been applied [37]. Apparently, the initial chloride gradient and thus the speed of diffusion of the residual chloride was reduced so strongly by chloride removal, that the distribution remained almost stationary. On the other hand, in reference slabs where the chloride had not been removed there was a distinct shift from the outer zone, where the chloride content was highest to the inner zones, which contained less chloride.

This example shows that the method promises long-term success in cases where excessive chloride concentrations are present mainly in the concrete cover, even if it has not been possible to reduce the chloride content to sub-critical values everywhere. If, however, the zone behind the outer reinforcement, where chloride removal is inefficient, is heavily contaminated, a renewed rise in chloride content is to be expected at the reinforcement. This is aggravated by the fact that the deeper zones do not dry out as easily, so that transport by diffusion is more rapid than in the outer zones of the concrete which are more likely to dry out. Polder *et al.* determined the chloride distribution after CE treatment of a set of prisms [30,53]. Other prisms were exposed outdoors for one year, not sheltered from rain, and chloride profiles determined [66]. A slight but significant redistribution of chloride was found. By finite element calculations it was shown that the rate of chloride redistribution agreed well with transport by diffusion. The diffusion coefficients for this redistribution were similar to those observed for chloride penetration during 16 years of immersion in the North Sea of the same material [30,53,67,68]. Apparently the concrete remained quite wet during the outdoor exposure. This was confirmed by relatively constant resistivities. Accordingly, in two OPC concretes with w/c 0.4 and 0.54, the chloride diffusion coefficients $D(Cl)$ were about 2×10^{-12} and 3×10^{-12} m^2/s respectively, and in a blast furnace slag cement (CEM III/B LH HS) concrete with w/c 0.4, $D(Cl)$ was 0.3×10^{-12} m^2/s.

The redistribution of any remaining chloride may cause the onset of corrosion. The time scale depends on the diffusion coefficient and the geometry (cover depth, steel reinforcement mesh size, remaining chloride content). In OPC this may occur within about ten years [66]. In concrete that dries out significantly, the situation is different. From resistivity measurements and assuming an inverse relationship between resistivity and diffusion [54,69], transport coefficients for semi-dry concrete may be estimated. In general, the resistivity of various concretes is about two to three times higher in equilibrium with 80% R.H. (average outdoor humidity in sheltered, West-European climate) than in water saturated conditions [70]. Accordingly, chloride redistribution after CE may be expected to be two to three times slower in semi-dry concrete than in wet (water saturated) concrete. In considering long-term durability it has to be mentioned that due to the increase in OH$^-$ content at the rebars the situation becomes much more favourable than would be the case with reduction of Cl$^-$ alone. The field tests/applications followed by Manning [44] and by Elsener *et al.* [71,72] showed that the treated areas — after application of a cementitious coating to avoid further ingress of chlorides — have so far remained in the passive state (as determined by half-cell potential measurements) for more than eight and six years respectively.

6
Side Effects

As discussed in Sections 4.4. and 5.2., respectively, electrochemical realkalisation and electrochemical chloride extraction may have several potentially detrimental side-effects, such as

- the alkali-silica reaction
- reaction reduction of bond between steel and concrete
- hydrogen evolution and embrittlement of reinforcing steel.

As electrochemical realkalisation is conducted for shorter treatment times and lower current densities than electrochemical chloride extraction some of the possible side effects are more relevant for the latter. Therefore, most studies on these effects have been focused on CE, but the conclusions are also valid for electrochemical realkalisation. In particular, the alkali silica reaction is a potential risk as the electrolyte most widely used for realkalisation is Na_2CO_3.

6.1. Alkali Silica Reaction (ASR)

An increase in OH^- concentration of the pore solution around the reinforcement has a beneficial effect in terms of corrosion protection. But in the case of concrete that contains reactive siliceous aggregate particles, this can be dangerous because it can initiate or accelerate the damage resulting from the ASR. To study this question, Bennett *et al.* conducted tests on two-month old concretes prepared with the same cement (type 1, according to ASTM C 150) and a chloride content of approx. 3.5 kg Cl/m^3 concrete. The concrete contained different aggregates, namely inert quartz sand, reactive chert, or reactive opal. The test was run at a slightly higher current density (approx. 6 A/m^2) and over a longer period (total charge up to approx. 3000 Ah/m^2) and with a sodium borate solution of higher concentration (0.3M) as outer electrolyte than that recommended by the authors for practical application ($< 5 A/m^2$; $< 1500 Ah/m^2$; 0.2M sodium borate solution) [37]. Under these more severe test conditions the opal-containing concrete suffered the most severe damage, developing cracks and alkali-silica gel in large quantities, while there was little damage to the control sample through which no current had passed. The use of a 0.1M lithium borate solution as outer electrolyte could prevent ASR by the formation of insoluble lithium silicate, which has a very low propensity to expand [24].

Al-Kadhimi and Banfill [23] found that electrochemical realkalisation of carbonated specimens reduces ASR compared to uncarbonated control samples. This is explained by the absence of calcium hydroxide in carbonated concrete acting as a reserve of hydroxide ions feeding the formation of an alkali-silica gel having exceptional swelling capacity. The differences between this work and others which show that CE increases ASR are traced back to the presence of calcium hydroxide in chloride contaminated concrete which provides high pH-values. In fact, the formation of an alkali-silica gel is only possible if the hydroxide ion concentration exceeds a threshold of 250–500 mM/L.

Based on concrete samples prepared without chloride into which steel cathodes had been embedded, Page *et al.* found that ASR can be triggered by redistribution of alkalis induced by current flow [24,25]. The concrete in this work was made from a combination of aggregates which included a proportion of potentially reactive siliceous material, but the bulk alkali content was controlled at a level below that which was known to be required to induce significant expansion from the alkali silica reaction when stored in a normal humid environment. The samples were exposed to different current densities for different lengths of time (0, 1, 3, 5, and 8 A/m^2 for 4, 8, and 12 weeks), and then the changes of sample lengths were measured over a period of up to one year. With the exception of the reference sample and the sample that had been exposed to 1 A/m^2 for 4 weeks only, all samples showed marked elongations and local cracks in the vicinity of the steel cathode after some months of storage. This was attributed to the accumulation, caused by the current flow, of alkalis at the cathode, where the initially sub-critical alkali hydroxide content became supercritical. However, no simple relation could be established between the extent of elongation and the charge quantities passed through the sample. At a charge density of about $1–2 \times 10^7$ Coulomb/m^2 (approx. 2700 to 5500 Ah/m^2) the samples experienced the largest expansions, whereas these were less at charge densities that were higher and lower than this range. According to the authors, this maximum can probably be explained by the dependence of the volume increase on the ratio between Na_2O and SiO_2 of the ASR-gel that has formed. Although the content of reactive SiO_2 was constant in the concrete the alkali content increased at the cathode while current passed through the concrete. Thus, ASR-gels with different $Na_2O : SiO_2$ ratios, and therefore with different swelling properties, developed depending on the duration of current flow. On the prevention of ASR by the use of lithium ions the authors report that the effect has been proven, but that caution is advised because inadequate lithium concentrations can make the expansion of concrete by ASR even worse [24]. Whether and how these diffculties in the case of lithium transport from the outer electrolyte can be counteracted is still being investigated.

Page *et al.* [25] also studied the changes in the composition of the pore solution at the cathode caused by current flow by using hardened cement paste samples containing 1% chloride (NaCl) and found an increase in OH$^-$ concentration at the cathode from about 0.5 to more than 2 mole/L, which is in agreement with other pore solution studies [58]. Such high OH$^-$ concentrations never occur in normal concrete, and the investigations of Page *et al.* show that aggregates that have been considered harmless so far may become reactive under such extreme conditions. To avert the danger of alkali-silica reaction and some other problems it has been proposed recently not to use the reinforcement as cathodes, but to place electrodes into bore holes. This would not only facilitate the removal of chloride from concrete but of alkalis as well and, in addition, prevent the formation of hydrogen at the reinforcement [57]. However, because of the risk of stray corrosion of rebars this technique seems to be not advisable.

6.2. Reduction of Bond Between Steel and Concrete

6.2.1. Influence of test conditions

It is known from studies carried out in the context of cathodic protection that the bond strength between the steel reinforcement and concrete can be impaired by current flow. The changes in the bond strength between steel and concrete occurring during electrochemical chloride removal were investigated by Bennett *et al.* [37]. They conducted pull-out tests on concrete prisms which contained centrally a longitudinal reinforcement bar with both ends protruding from the sample. They measured the tensile forces required to move the steel by 0.25 mm at the loaded end and by 0.025 mm at the free end from the prism. They also determined the maximum force needed to disintegrate completely the bond between steel and concrete (ultimate bond stress). Twenty-eight of the 30 specimens had been prepared with the addition of chloride (4.5 kg/m^3, Cl$^-$/concrete), two samples were free from chloride. The chloride-containing samples were exposed to different current densities (0.02, 1, and 5 A/m^2), and various total charge quantities were passed through the concrete varied (200, 500, and 2000 Ah/m^2). The chloride-free samples were not polarised. It was found that the ultimate bond stress was reduced by the current flow in the chloride-containing prism on average by 11% without any noticeable effect of current density or charge quantity.

The force required to move the steel by 0.25 mm (0.025 mm) at the loaded (unloaded) end was about 40% (about 60%) less if either the current density or the total charge or both were high. However, the authors pointed out that further statistical analyses showed that the reduction in the tensile force needed at the loaded steel end of chloride-containing samples was significant only when both the current density and the total charge quantity were high (5 A/m^2 and 2000 Ah/m^2 respectively). At the free end, on the other hand, significant reductions in the necessary tensile force were observed when either the current density or the charge quantity passing through the test specimen was high. Thus, the test results did not provide an answer to the question whether current density or the charge quantity is the more important factor in the reduction of bond strength.

In the chloride-free samples the ultimate bond stress was only slightly smaller than in the chloride-containing reference sample through which also no current had passed; but the extraction forces measured at the loaded end and at the free end were lower in the chloride-free sample than in all samples with current flow. According to the authors the reasons for this are unknown, but they suggest that corrosion products evolved in the chloride-containing specimens may be responsible for the increase in bond strength.

Ihekwaba *et al.* demonstrated that application of electrochemical chloride extraction to reinforced concrete alters the pull-out strength and bond between the embedded steel and surrounding concrete [73]. The specimens in this study containing chloride ion concentrations of 1.7% and 3.0% by weight of cement respectively were electrochemically treated with two current densities of 1 and 3.0 A/m^2 of concrete surface (9 and 28.6 A/m^2 of the exposed steel surface) for eight weeks. CE treated specimens showed significant reductions in pull-out strength, with the degradation in bond being dependent on the applied current density and initial chloride contamination. The alkali ion accumulation around the steel rebars was observed to follow a similar trend to that of bond degradation. This accumulation appears to be detrimental to the interface integrity and can lead to disbondment. Results of pull-out tests after different electrochemical chloride removal treatments (current densities were 1.6, 4 and 8 A/m² related to rebar surface and the durations of the treatment were 7, 14, 26 and 56 days) have also shown a significant reduction in bond stress; but this was observed only within the charge range 600– 5000 Ah/m² of rebar steel [74,75]. At higher charges, however, there was an extreme increase in the bond stress. The reference specimen which was not treated electrochemically, also contained chlorides. The authors mentioned that their results were somewhat contradictory, which could be caused by the values being liable to a certain degree of uncertainty since they were based on only a single measurement.

Nustad published a review of the results from different investigations on the possible changes in the steel-to-concrete bond strength as a result of desalination [76]. Although most of the findings were carried out at a current density of 1 A/m² of concrete surface, the steel-to-concrete ratio of the specimen was often far lower than normal for real structures, resulting in quite abnormal current densities at the rebar surface which are irrelevant for ordinary chloride removal.

Nustad comes to the conclusion that the chloride content of the specimens tested as well as the extent of rebar corrosion of the control specimens obviously play an important role [76]. It seems clear that an enhanced bond strength of salty control specimens must be expected due to reinforcement corrosion. The explanation for this is that the formation of corrosion product on specimens containing chlorides 'prestresses' the concrete and increases the rebar pull-out strength of chloride-containing vs chloride-free specimens. During the electrochemical treatment the reinforcement is cathodically polarised which results in chloride ions being repelled from the rebar and any corrosion products being dissolved. Subsequently, any enhanced bond strength as a result of rebar corrosion must be expected to disappear. In order to evaluate any changes in bond strength as a result of electrochemical treatments it is therefore essential that the treated specimens are compared with chloride-free control specimens cured and stored under comparable conditions.

Broomfield also mentioned that the comparison of cathodically polarised specimens with corroding samples can be misleading [77]. He discusses the results of bond strength tests where the bond strength of the chloride-containing specimens was around 57% higher than those of the chloride-free specimens. At no time did the bond strength of specimens containing chloride fall below the bond strength of the equivalent (i.e. same age as CE) non-chloride specimens prior to CE.

6.2.2. Causes

The decrease in bond strength is caused by changes in the composition of the hardened cement matrix that occur in the direct voltage field, such as the significant accumulation of alkali hydroxide around the cathode. This may lead to a softening up of the binder matrix and thus to a weakening of the bond. Page *et al.* have observed that the sulfate (SO_4^{2-}) concentration of the pore solution and the ratio of free/bound Cl^- rise around the cathode during current flow (current density max. 5 A/m²; max. duration of test 12 weeks), from which they conclude that the stability of the sulfo- and chloro-aluminate compounds is weakened by the high local concentration of alkalis. They suggest also that the composition of the C-S-H-phase is modified by the entry of Na^+ and K^+. Micro hardness tests performed on the hardened cement paste samples near the cathode did not provide any statistically proven information on changes in strength, because of the considerable scatter in the results [25,27].

Bennett *et al.* also noted a disintegration of the sulfur phases in concrete samples caused by rising alkali hydroxide concentration [37]. However, their specimens were exposed to much higher current densities (20 A/m^2; 5000 Ah/m^2) than the samples investigated by Page *et al.* Electron microscope studies (element distribution analyses) revealed a marked increase in the total content of sodium and sulfur around the outer reinforcement. The authors found clear indications for deposits of sodium- and sulfur-rich compounds in the pores of the hardened cement matrix. They also found a significant increase in the porosity of the hardened cement matrix, with a much higher proportion of pores between 1 and 10 µm in diameter than in the non-polarised reference sample. Specimens tested at current densities of 6 and 2 A/m^2 respectively, and at a total charge of 3000 Ah/m^2 did not present any detrimental effects attributable to current flow. In one case the high current densities and charge quantities produced a crack at the level of the outer reinforcement parallel to the sample surface which caused a disintegration of the cover of the whole sample (concrete slab; 60×60 cm). This is attributed by the authors to hydrogen gas which had formed at the reinforcement apparently faster than it could escape from the concrete; the pressure that built up then caused the disintegration [37].

A small and probably statistically not significant increase of microcrack density was found in a preliminary thin section microscopy study by Polder of desalinated concrete prisms at 1 and 4 A/m^2 steel surface for 39 days [30,53]. No changes in the concrete microstructure were observed.

6.3. Hydrogen Evolution and Embrittlement of Reinforcing Steel

Negative (cathodic) polarisation results in hydrogen being evolved at the steel surface. This may have adverse effects because of the pressure exerted by the gas on the concrete and the danger of reinforcement embrittlement. As explained in Section 5.2., gas pressure appears to be a problem only at current densities and total charges far above those applied in practical cases. However, hydrogen evolution may induce hydrogen embrittlement of prestressed steel. There is considerable debate on this matter (see Section 7.2.).

Bennett *et al.* examined whether hydrogen evolution is indeed harmless to ordinary reinforcement — as is generally believed [37]. They first subjected notched steel samples that had been immersed into an aerated, $Ca(OH)_2$ saturated solution and polarised cathodically to tensile tests at constant elongation speed. These showed that neither the current densities normally applied in chloride removal nor the chloride concentration of the $Ca(OH)_2$ solution had any significant adverse influence on the ultimate breaking load of the steel. However, a significant reduction in breaking elongation of up to 80% compared to the non-polarised reference sample was observed. So, while the current flow had no influence on strength, it did significantly influence the plastic properties of the reinforcing steel. Apparently sufficient hydrogen had been absorbed to change the mechanical properties, at least in the short term.

However, further tests showed that elongation was restored quickly to about 90% of the initial value after the current flow had been cut off suggesting quick release of hydrogen from the steel.

Similar tests were also conducted on mortar specimens with steel samples that were not notched but which contained a zone of reduced cross section. The whole part that included the reduced cross section was embedded in mortar, and the two ends protruded from the mortar. The results corresponded qualitatively to those obtained from the notched samples. The authors concluded that the current densities and charge quantities recommended for chloride removal do not have an adverse impact on ordinary reinforcement.

Contrary to ordinary reinforcement, high strength prestressing steels are generally sensitive to hydrogen-induced stress corrosion cracking. This type of cracking is caused when hydrogen infiltration leads to an embrittlement of the material. The process of cracking is a combination of cathodic hydrogen evolution and physical transport and fracture processes inside the steel. For high-strength steels this type of damage requires only a very small amount of atomic hydrogen capable of absorption. The critical hydrogen concentration depends on the type of prestressing steel, its grain structure and the stress level. On post-tensioned structures, where the high strength steel is in a grouted metallic or plastics duct, different opinions exist on the risk for hydrogen embrittlement during the application of CE or RA. Further research work is needed in order to substantiate the claim that a metallic duct can have a beneficial effect by acting as a Faraday cage and shielding the high strength steel in the duct from the cathodic current [19].

6.4. Other Concrete Properties

Experimental investigations into effects of electrochemical realkalisation on concrete have shown no detrimental effects on the concrete properties [22]. They change in a manner consistent with the deposition of material (i.e. reaction products of the electrolyte — penetrating from outside — with the concrete) in the pores of the concrete leading to densification. The pore size distribution changes in the direction of smaller pores. The total water absorption, capillary absorption and initial surface absorption all decrease. The compression strength, flexural strength, pull-out strength, dynamic modulus of elasticity and ultrasonic pulse velocity all increase. The water absorption and capillary absorption results, combined with visual observations of the cut specimens provide no evidence in support of the assertion that realkalisation produces a network of fine cracks throughout the treated concrete.

Scanning electron miscroscope (SEM) investigations of realkalised specimens both from laboratory tests [9] as well as from field experiments [14] have shown no differences compared to non-treated specimens.

7

Aspects of Practical Application

7.1. Feasibility

In general, both electrochemical methods, i.e. realkalisation and chloride removal, are believed to be most suitable for cases where no, or only limited, damage to concrete is present. This means that the feasibility is good for preventing corrosion in structures where the concrete cover has been contaminated with chloride but the threshold concentration has not yet been exceeded or the carbonation front has not yet reached the level of the reinforcement and corrosion has consequently not yet initiated. Furthermore, for economic reasons these electrochemical techniques may also be feasible if only limited damage is present (high cost of necessary conventional repair). As several field tests have shown, the treatments can be applied successfully also when the chloride threshold is exceeded and even when corrosion has already started. The only conditions are the structural safety (reduction in cross section) and spalling of the concrete (economic reasons). In general, the techniques are more cost effective if applied before deterioration has become so severe as to require extensive removal and replacement of spalled and delaminated concrete (conventional repairs).

As the current flow reduces the plastic properties of the reinforcing steel (see 6.3.) dynamic loading of the structure during the treatment could be detrimental. Therefore, it is advisable to consider measures which ensure that the loadings during the treatment are below maximum permissible values (e.g. by weight or speed limitations — where road surfaces are involved).

In cases where deterioration of a structure depends on the rate of chloride penetration from the environment and on the maintenance period of the treated structure, the concrete surface may be sealed against further chloride ingress, or the treatment may need to be repeated periodically. If the source of chlorides has not been eliminated it is recommended that a cementitious coating be applied.

In order to make both techniques work properly, several requirements have to be met and the structure must be thoroughly investigated before their application.

7.2. Preliminary Investigations [61]

7.2.1. General

The assessment of the structure should include a general survey, identifying the presence of structural cracks, deformations and other obvious defects. If such defects are present to a significant extent, the treatment should be reconsidered and structural measures should be taken first. If not, the inspection should focus on the preparation for the electrochemical treatment. The following items should be measured over representative parts of all areas (or types of concrete components or structures) to be treated.

7.2.1.1. Concrete cover to the steel
This includes measuring the cover depth, its average and variation. It should be noted that a large variation in cover will cause the chloride removal efficiency or the realkalisation effect to be lower than for a small variation. If cover variation is very large, it may be impracticable to apply these techniques.

7.2.1.2. Chloride content and distribution
Prior to an electrochemical chloride removal treatment the chloride profile, lateral variations and the source of the chloride should be determined. In particular establishing the chloride content and its distribution is important as the effect of the treatment is evaluated from these parameters. Identifying locations for process control cores at this stage is advisable. For geometrical reasons it is more difficult to remove mixed-in chloride added at the time of mixing than it is to remove penetrated chloride [48]; in general removal of mixed-in chloride is less likely to be successful.

7.2.1.3. Carbonation depth
Prior to an electrochemical realkalisation treatment the carbonation depth should be measured at representative locations. Identifying locations for process control cores at this stage is advisable.

7.2.1.4. Corrosion state of the rebars
For both electrochemical techniques (RA and CE) a potential mapping should be performed prior to the treatment to get a precise idea of the corroding zones and to know the initial potential field. The upper map of Fig. 18 gives an example of a potential field measured on a sidewall before chloride removal.(A colour plate of this figure can be found at the end of the book, facing p. 58.)

7.2.1.5. Electrical continuity of the reinforcement
The reinforcing steel should be electrically continuous for successful electrochemical treatment. Realkalisation due to electrolysis is impossible and chloride will not be removed efficiently from the vicinity of the bars that are not connected to the negative terminal of the current source. Discontinuities have to be corrected by providing additional connections. If a structure contains many discontinuous rebars, it may be impracticable to apply electrochemical rehabilitation techniques. It is emphasised, however, that steel discontinuity is not met very often in practice.

7.2.1.6. Electrical continuity of the concrete
The concrete around the steel and between the steel and the anode has to provide continuous electrolytic conduction. That means that the concrete should not contain major corrosion cracks, delaminations or old repairs with a high electrical resistivity (non-cementitious polymer mortar repairs or coatings) because these may hinder a uniform current flow.

7.2.1.7. The presence of potentially alkali-reactive aggregates
Because these electrochemical treatments will increase the alkali content around the rebar, local ASR expansion may be stimulated [24,26] (see also 6.1.). The reactivity of

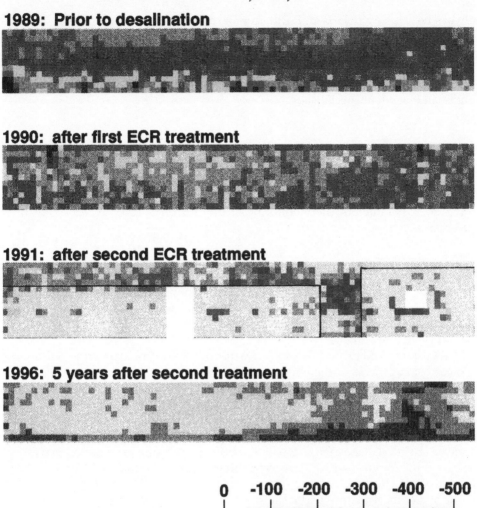

Fig. 18 *Potential field measured on a sidewall before chloride removal, after first and second treatments and six years after the treatments. Potential values are related to a Cu/CuSO$_4$ reference electrode* [14].*(A colour plate of this figure can be found at the end of the book, facing p. 58.)*

the aggregate should be checked for by thin section microscopy; if it is potentially reactive at least a trial is advised.

7.2.1.8. The presence of prestressing steel (see 6.3.)
With strong negative polarisation, hydrogen is evolved at the steel surface and in the case of prestressing steel may induce hydrogen embrittlement. There is considerable debate on this item. Post-tensioned steel in perfect ducts will be shielded from strong polarisation but in practical structures, defects in the ducts may occur and the risk may be real. Pretensioned steel in direct contact with the concrete will

always be strongly polarised. Furthermore, some types of steel are more sensitive than others. Because of the high risk of structural damage due to hydrogen embrittlement, application of electrochemical realkalisation or electrochemical chloride extraction to prestressed structures is not recommended unless trials have shown that no damage to the prestressing steel can occur. Such trials should be carried out by independent specialists.

7.3. Design of Treatment [61]

7.3.1. General

The design of the electrochemical treatment involves the choice of the anode system, the process parameters, the monitoring of the treatment and the criteria for final acceptance.

7.3.1.1. Anode system

The anode system consists of an anode, a current distributor system and the electrolyte. The anode may be a mesh of steel or an inert material such as activated titanium. For non-inert metal anodes oxidation has to be considered as an anodic reaction; rust staining of the concrete surface from the steel mesh anode is of practical relevance.

The current is fed to the anode by wires that must be adequately protected against the corrosion process at the anode. It is important that in any case short circuits between anode and reinforcing steel be avoided. The anode is surrounded by the electrolyte which in the case of realkalisation is usually sodium carbonate solution, and in the case of chloride extraction is usually saturated lime solution or tap water. The electrolyte liquid is usually mixed with fibres to form a paste and subsequently sprayed onto the concrete surface, or it is contained in a temporary shuttering (vertical surfaces) or troughs (soffits), which together are termed 'tank' or 'containerised' applications.

Normally the structure surface is split into different zones (similiar to cathodic protection) for the treatment. Each zone has its own separate anode and current feed unit. There is a tendency towards using smaller zones (from hundreds of square meters down to tens of square meters), which is regarded as beneficial because in such smaller zones better control and more uniform current distribution are possible.

7.3.1.2. Process parameters

The process parameters are current density and treatment time and for both types of electrochemical treatments it is important that these are adjusted with respect to the local conditions and requirements.

7.3.1.3. Electrochemical realkalisation

The amount of hydroxide produced at the reinforcement is determined by the total charge that has passed between the steel and the anode. The amount of charge is the integral of current multiplied by the time. A higher current density in a shorter period of time will in principle generate the same amount of hydroxide as a lower current density during a longer time, provided that the total charge is equal. Current independent transport processes depend mainly on carbonation depth, concrete quality and humidity. The design current density is typically in the range of 0.8–2 A/m² concrete surface area. An upper limit has to be specified to avoid deterioration of the concrete. The time required is typically between 3 and 14 days [11,16,78]; it seems to depend on:

- the carbonation depth
- the quality of concrete
- the concrete cover thickness
- the distribution of steel reinforcement
- the distribution of the current; if the current distribution is very inhomogeneous, longer treatment times are necessary.

7.3.1.4. Electrochemical chloride extraction

The amount of chloride removed is determined by the total charge that has passed between the steel and the anode and by the chloride removal effciency.

The design current density is typically in the range of $0.8 - 2$ A/m² concrete surface area. An upper limit has to be specified to avoid deterioration of the concrete. Upper limit values of 2 to 5 A/m² are given in the literature [37,44,48].

The removal efficiency has been reported from laboratory studies to be between 0.1 and 0.5. In practical cases it is probably between 0.1 and 0.3 [37]. This means that only between 10% and 30% of the total current is actually removing chloride ions. The remaining part of the current is transported by ions other than chloride (hydroxide, alkalis, etc.).

The time required is typically between a few weeks and a few months and seems to depend on:

- the initial chloride content of the concrete
- the chloride distribution within the concrete
- the source of chloride (see 7.2.1.2.)
- the quality of concrete
- the distribution of steel reinforcement
- the distribution of the current; if the current distribution is very inhomogeneous, a lower overall chloride removal efficiency must be expected.

7.3.1.5. Monitoring

During the treatment, some form of monitoring of progress is necessary. For RA simple phenolphthalein tests on drilled cores may be used for monitoring of the progress. For documentation the sodium uptake should be measured. In the case of CE the remaining chlorides in drilled cores are measured. For liquid electrolytes the progress may be monitored by analysing the chloride content of the electrolyte. Provided the electrolyte pH is high enough to prevent chlorine gas evolution, the amount of chloride in the electrolyte is equal to the amount of chloride removed from the concrete. Under that condition, the chloride in the electrolyte may be used to monitor the progress [30,54]. Close monitoring of current density and driving voltage development may also help to determine the progress of the treatment.

7.3.1.6. Acceptance criterion
The acceptance criterion should be specified in the design stage. In the case of realkalisation this is normally a minimum of introduced alkalis and/or the proof of the repassivation of the steel surface. In the case of chloride extraction it is normally a maximum allowed average chloride content remaining after the CE treatment. It is advisable to specify clearly where cores should be taken, what analysis technique should be used and how the 'average' should be calculated. It should be realised that the chloride can be removed quite easily from the concrete around the rebars, but with more difficulty from the concrete between and behind rebars.

7.3.1.7. Additional protection
In the design stage the need for an additional protective barrier after the treatment should be specified, for instance the application of an organic coating, a silane/siloxane hydrophobic treatment or a cementitious coating.

7.4. Carrying out the Treatment

Before the treatment, the concrete surface is cleaned, usually by grit blasting or water cleaning, in order to remove dust, oil and grease, which may hinder the current flow. In addition, it may be necessary to seal open cracks to prevent the development of low resistance paths to the reinforcement which decrease the efficiency. Concrete which has spalled or delaminated due to reinforcement corrosion must be replaced by appropriate repair material (electrically compatible which, in practice, means cementitious).

During the treatment, the current must be applied in a controlled way and as close as possible to the design value. The progress of the process must be monitored by one of the methods described above. A fibre electrolyte should be wetted regularly to prevent high resistance buildup. Liquid in tanks should circulate to keep the composition constant.

After the treatment the anode system is removed and the concrete surface is cleaned. If a steel mesh has been used as anode, the surface may be stained by rust, which must be removed by light grit or water blasting.

Finally, the protective layer barrier must be applied if specified.

7.5. Final Acceptance

No official standards exist on acceptance of these techniques.

7.5.1. Electrochemical realkalisation

For electrochemical realkalisation the final acceptance will depend on the achievement of a repassivated steel surface and/or a sufficient increase of the pH-value in the pore water. Alkali analyses are performed on drilled cores on sections parallel to the surface to obtain a profile, or *en bloc* to determine the total amount of sodium in the concrete cover. In the state of equilibrium with a constant carbon dioxide concentration of the atmosphere the pH-value of the pore water will be given by the ratio between sodium carbonate and sodium hydrogen carbonate and this is controlled by the amount of sodium, i.e. sodium analyses can be used as an indicator for the pH when chemical equilibrium has been established [11,12]. Potential measurements [12,18,19] or anodic polarisation of the steel reinforcement are further tools to assess the degree of realkalisation of carbonated concrete [6,9,10]. However, potential mapping is not recommended without calibration for the actual structure, because the introduced sodium carbonate changes the resistivity of the concrete [12]. Thus, it has to be appreciated that after electrochemical realkalisation the potentials shift to values that are more negative than before the treatment. This behaviour can be rationalised by taking into account that (a) the concrete after the realkalisation has a much lower resistivity because of the sodium carbonate uptake mentioned above, and (b) passive steel in concrete acts as a pH electrode which in this case means lower potentials due to the higher pH value after the treatment. The presence of a homogeneous potential field after the treatment is considered to be a strong indication of repassivation [19].

7.5.2. Electrochemical chloride extraction

In the case of chloride removal the final acceptance will depend on the achievement of a sufficiently low chloride content, preferably below the corrosion initiating threshold. Potential mapping (and other corrosion monitoring techniques) before and after the treatment will give valuable additional information. For obtaining representative potentials "after", they must have relaxed from the strong polarisation produced by the treatment; in practice this may take weeks or even months [50]. Potential mapping can also be used to monitor the long term behaviour (see Fig. 18).

7.6. Maintenance

In principle, no maintenance is necessary. It is advisable to monitor the penetration of chloride and/or corrosion initiation over the rest of the service life after the treatment. This can be obtained by regular inspections (e.g. every five years) or by using embedded monitoring sensors.

8
Conclusions

8.1. General

Electrochemical realkalisation (RA) and electrochemical chloride extraction (CE) lead to an increase in pH at the rebars and to repassivation of corroding steel. The durability of CE has been proven on different site jobs with a track record of more than five years during which time further chloride ingress has been avoided by applying a coating on the concrete surface. Several reinforced concrete structures treated with RA have shown good performance over several years without applying a coating.

Electrochemical realkalisation and chloride extraction should not be applied to structures suffering corrosion resulting from very low cover or very low concrete quality. In those cases the required control of the current density and current distribution cannot be achieved. Moreover, after the treatment early washing out of alkalinity or penetration of chloride will be probable.

Behind the reinforcement and also when the space between reinforcing bars is relatively large, shadow spots can develop since the current flow reaches mainly the first layer of reinforcement.

Except for any alkali-silica reaction, the side-effects described in Section 6 can be avoided or kept to such a low level by respecting the given constraints (low current densities, total charge not too high) that they will cause no damage. But if the concrete contains particles with reactive silica, there is a heightened probability of damage in the vicinity of the reinforcement because of rising alkali hydroxide concentrations. According to laboratory tests, this could be remedied by using a lithium hydroxide or lithium borate solution, respectively, as electrolyte, but this has not yet been sufficiently investigated in actual structures.

As the negative polarisation during these treatments causes hydrogen evolution the risk of hydrogen-induced embrittlement of prestressing steel cannot be completely excluded. Therefore, these techniques should not be applied to tensioned structures. For post-tensioned structures these techniques are not recommended unless investigations have shown that no damage can occur.

The decision for one of the electrochemical methods or a conventional repair strategy depends on the degree and cause of damage as well as the future lifetime and of course the required costs. Taking into account that for long-term durability of conventional repair often large areas of contaminated concrete have to be removed then electrochemical methods become competitive also from an economical point of view. Chloride removal of concrete structures manufactured with chloride-containing materials is less successful. For this case cathodic protection is advised.

As in practical cases any necessary repairs are often carried out only when the damage has become obvious. Therefore, the electrochemical methods may not always convince those responsible for the maintenance of construction, especially since tensioned concrete constructions must be excluded as candidates. Earlier checks would help to avoid conventional repairs (replacement of contaminated concrete), which are not only more expensive but also much more harmful to workers and environment. Earlier repairs would also have the additional benefit that future carbonation or ingress of chloride into the structure could usually be inhibited by preventive measures. As the electrochemical technique is a non-destructive method of repair, the monolithic concrete structure can be maintained which is usually not the case with conventional methods. This is a considerable advantage from the point of view of the static strength of the concrete, and may be essential even for certain elements, such as heavily loaded columns. In distinction to cathodic protection systems with a shotcrete overlay there is no weight added by the temporary realkalisation or chloride removal treatment.

8.2. Electrochemical Realkalisation

Steel surfaces activated due to carbonation of the surrounding concrete can be repassivated by a temporary cathodic polarisation of the reinforcing steel. The repassivation effect of electrochemical realkalisation has been confirmed in anodic polarisation tests. The success of the treatment is mainly determined by electrochemical reactions at the steel surface. Penetration of alkaline electrolyte from the concrete surface is controlled by diffusion and absorption, electroosmotic mechanisms are of minor importance.

From anodic polarisation tests it can be concluded that the phenolphthalein test is not a sufficient tool to prove the state of the steel surface, i.e. whether it is really repassivated or not. As the buffer capacity of calcium hydroxide in non-carbonated concrete cannot be re-established by an electrochemical method it is important that sufficient amounts of alkali penetrate into the treated structure.

Although there are open questions concerning criteria for assessing the success of the treatment as well as the long-term durability the process is considered of particular interest for the following structures:

- concrete facades which cannot or should not be changed
- silos and cooling towers with thin dimensions
- concrete structures where additional weight is not desired or not possible.

8.3. Electrochemical Chloride Extraction

The reports and publications on electrochemical chloride removal clearly indicate that a considerable amount of chloride can be extracted by electrochemical chloride extraction.

Main application examples for electrochemical chloride extraction are as follows:

- concrete structures contaminated by de-icing salt
- concrete structures along the coast
- concrete structures of the chemical industry
- concrete structures with chloride contamination due to PVC fires.

However, the efficiency of the method will be greatly limited in case of chloride that has penetrated far behind the reinforcement or present as mixed-in chlorides.

References

1. "Draft Recommendation for Repair Strategies for Concrete Structures Damaged by Reinforcement Corrosion", RILEM TC124 Ed. P. Schiessl, *Mater. Struct.*, 1994, **27**, 415–436.

2. K. Tuutti, Corrosion of Steel in Concrete, Swedish Cement Concrete Research Institute, Stockholm, CBI Research (1982), FO 4/82.

3. Ø. Vennesland and J. B. Miller, Norwegian Patent Application No. 875438, 1987.

4. J. B. Miller, European Patent Application No. 90108563.9, 1990.

5. EPO patent no. 0264421, "Method of Electrochemical Realkalisation of Concrete" (revoked in 1997).

6. G. Sergi, R. I. Walker and C. L. Page, Mechanisms and criteria for the realkalisation of concrete, in *Corrosion of Reinforcement in Concrete Construction*, Eds C. L. Page, P. B. Bamforth and J. W. Figg, Royal Society of Chemistry, Cambridge, 1996, pp.491–500.

7. A. J. van den Hondel and R. B. Polder, Electrochemical realkalisation and chloride removal of concrete, *Construction Repair*, September/October 1992, **19**, 20.

8. P. F. G. Banfill, Features of the mechanism of realkalisation and desalination treatments for reinforced concrete, in *Proc. Int. Conf. on Corrosion and Corrosion Protection of Steel in Concrete*, Ed. R. N. Swamy, Sheffield, 1994, pp.1489–1498.

9. J. Mietz, B. Isecke, B. Jonas and F. Zwiener, "Elektrochemische Realkalisierung zur Instandsetzung korrosionsgefährdeter Stahlbetonbauteile", Abschlußbericht zum BMFT-Forschungsvorhaben 03 F61 5A9, BAM, Berlin, Oktober 1994, "Electrochemical realkalisation for rehabilitation of reinforced concrete structures prone to corrosion", Final Report of the BMFT Research Project 03 F615A9, BAM, Berlin, October, 1994, (in German).

10. J. Mietz, Electrochemical realkalisation for rehabilitation of reinforced concrete surfaces, *Mater. Corros.*, 1995, **46**, 527–533.

11. J. A. Roti, "Elektrochemische Realkalisierung und Entsalzung von Beton", "Electrochemical realkalisation and desalination of concrete", WTA-Berichte, 1990, **6**, 131–147, (in German).

12. L. Odden, The repassivating effect of electrochemical realkalisation and chloride extraction, *Proc. Int. Conf. on Corrosion and Corrosion Protection of Steel in Concrete*, Ed. R. N. Swamy, Sheffield, 1994, pp.1473–1488.

13. B. Isecke and J. Mietz, Investigation on realkalisation of carbonated concrete, *Proc. EUROCORR '91*, Budapest, 1991, Vol. II, pp.732–738.

14. B. Elsener and H. Böhni, "Elektrochemische Instandsetzungsverfahren — Fortschritte und neue Erkenntnisse", "Electrochemical rehabilitation methods — Progress and new results", Referate der Tagung "Erhaltung von Brücken — Aktuelle Forschungsergebnisse", "Maintenance of bridges — Actual research results", SIA Documentation D 0129, Schweizer Ingenieur- und Architektenverein, Zurich, 1996, pp.47–59, (in German).

15. P. Haardt and H. K. Hilsdorf, "Realkalisierung karbonatisierter Betonrandzonen durch großflächigen Auftrag zementgebundener Reparaturschichten", "Realkalisation of carbonated surface zones by means of cement-based repair layers", 4. Kolloquium "Werkstoffwissenschaften und Bausanierung", 4th Colloquium "Materials Science and Restoration" TA Esslingen, 1992 (in German).

16. J. A. Roti, "Betoninstandsetzung mittels elektrochemischer Realkalisierung und Chloridentfernung", "Concrete repair by means of electrochemical realkalisation and chloride removal", 2. Int. Kongreß zur Bauwerkserhaltung, *2nd Int. Congress on Maintenance of Structures*, 1994, Berlin (in German).

17. D. Jungwirth, P. Grübl and A. Windisch, "Elektrochemische Schutzverfahren für bewehrte Bauteile aus baupraktischer Sicht", Electrochemical protection measures for reinforced structures from the view of practice, *Beton Stahlbetonbau*, 1991, **86**, 190–192 (in German).

18. B. Elsener, R. Gabriel, L. Zimmerman, D. Bürchler and H. Böhni, Electrochemical realkalisation – Field experience, Extended Abstracts of *EUROCORR '96*, Session II-OR 17-1 – 17-4. Publ. CEFRACOR and Soc. Chim. Ind., Paris, 1996.

19. B. Elsener, L. Zimmerman, D. Bürchler and H. Böhni, Repair of reinforced concrete structures by electrochemical techniques – Field experience, Extended Abstracts of *EUROCORR '97*, Vol. I, pp.517–522. See also *Corrosion of Reinforcement in Concrete — Monitoring, Prevention and Rehabilitation*, Eds J. Meitz, B. Elsener and R. Polder. EFC Publication No. 25, published by The Institute of Materials, London, 1998, pp.125–140.

20. J. S. Mattila, M. J. Pentti and T. A. Raiski, Durability of electrochemically realkalised concrete structures, in *Corrosion of Reinforcement in Concrete Construction*, Eds C. L. Page, P. B. Bamforth and J. W. Figg, Royal Society of Chemistry, Cambridge, 1996, pp.481–490.

21. J. Mietz, B. Jonas and F. Zwiener, "Die elektrochemische Realkalisierung carbonatisierten Betons", "Electrochemical realkalisation of carbonated concrete", Berichtsband über das 33. Forschungskolloquium des Deutschen Ausschusses für Stahlbeton (Berlin 1996), Hrsg. von der BAM, Abt. VII, 1996, pp.15–19 (in German).

22. T. K. H. Al-Kadhimi, P. F. G. Banfill, S. G. Millard and J. H. Bungey, An experimental investigation into effects of electrochemical realkalisation on concrete, in *Corrosion of Reinforcement in Concrete Construction*, Eds C. L. Page, P. B. Bamforth and J. W. Figg, Royal Society of Chemistry, Cambridge, 1996, pp.501–511.

23. T. K. H. Al-Kadhimi and P. F. G. Banfill, The effect of electrochemical realkalisation on alkali-silica expansion in concrete, *Proc. 10th AAR Conf.*, Melbourne, August, 1996.

24. C. L. Page and S. W. Yu, Potential effects of electrochemical desalination of concrete on alkali silica reaction, *Mag. Concr. Res.*, 1995, **47**, (170), 23–31.

25. C. L. Page, S. W. Yu and L. Bertolini, Some potential side-effects of electrochemical chloride removal from reinforced concrete, *UK Corrosion & EUROCORR '94*, Bournemouth, UK, 1994, pp.228–238. The Institute of Materials, London, 1994.

26. G. Sergi, C. L. Page and D. M. Thompson, Electrochemical induction of alkali-silica reaction in concrete, *Mater. Struct.*, 1991, **24**, 359–361.

27. L. Bertolini, S. W. Yu and C. L. Page, Effects of electrochemical chloride extraction on chemical and mechanical properties of hydrated cement paste, *Adv. Cem. Res.*, 1996, **8**, (31), 93–100.

28. C. Andrade, M. A. Sanjuan, A. Recuero and O. Rio, Calculation of chloride diffusivity in concrete from migration experiments, in non steady-state conditions, *Cem. Concr. Res.*, 1994, **24**, (7), 1214–1228.

29. B. Elsener, "Ionenmigration und elektrische Leitfähigkeit im Beton", in *Proc. Symp. Elektrochemische Schutzverfahren für Stahlbetonbauwerke*, SIA Documentation D 065, Schweizer Ingenieur- und Architektenverein, Zürich, 1990, 51–59. B. Elsener, Ion Migration and Electrical Conductivity in Concrete, in *Proc. Symp. Electrochemical Protection Measures for Reinforced Concrete Structures*, SIA Documentation D 065, Schweizer Ingenieur- und Architektenverein, Zurich, 1990, pp.51–59 (in German).

30. R. Polder, Electrochemical chloride removal of reinforced concrete prisms containing chloride from sea water exposure, in *UK Corrosion & EUROCORR '94*, 1994, pp.239–248. Publ. The Institute of Materials, London, 1994.

31. D. R. Lankard, J. E. Slater, W. A. Hedden and D. E. Niesz, "Neutralisation of Chloride in Concrete", Report No. FHWA-RD-76-60, 1975, pp.1–143.

32. J. E. Slater, D. R. Lankard and P. L. Moreland, "Electrochemical Removal of Chlorides from Concrete Bridge Decks", Transportation Research Record No. 604, 1976, pp.6–15.

33. G. L. Morrison, Y. P. Virmani, F. W. Stratton and W. J. Gilliland, "Chloride Removal and Monomer Impregnation of Bridge Deck Concrete by Electro-Osmosis", Report No. FHWA-KS-RD-74-1, 1976, pp.1–41.

34. Noteby, European Patent Application No. 86302888.2, 1986.

35. Norwegian Concrete Technology: Personal information, 1994.

36. J. B. Miller, European Patent Application No. 90108563.9, 1990.

37. J. Bennett, T. J. Schue, K. C. Clear, D. L. Lankard, W. H. Hartt and W. J. Swiat, "Protection of Concrete Bridge Components: Field Trials", Strategic Highway Research Program, Report SHRP-S-657, 1993, 201.

38. J. Bennett, K. F. Fong and T. J. Schue, "Electrochemical Chloride Removal and Protection of Concrete Bridge Components: Field Trials", Strategic Highway Research Program, Report SHRP-S-669, 1993, 149.

39. J. Bennett and T. J. Schue, "Chloride Removal Implementation Guide", Strategic Highway Research Program Report SHRP-S-347, 1993, 45.

40. J. Bennett and T. J. Schue, "Evaluation of NORCURE Process for Electrochemical Chloride Removal from Steel-Reinforced Concrete Bridge Components", Strategic Highway Research Program, Report SHRP-C-620, 1993, 31.

41. S. J. Pate, Chloride extraction on the midland links viaducts, *Constr. Rep.*, 1996, July/August, 16–19.

42. K. Armstrong, M. G. Grantham and B. McFarland, The trial repair of Victoria pier, St. Helier, Jersey using electrochemical desalination, in *Corrosion of Reinforcement in Concrete Construction*, Eds C. L. Page, P. B. Bamforth and J. W. Figg. Royal Society of Chemistry, Cambridge, 1996, pp.466–477.

43. J. B. Miller, Chloride removal and protection of reinforced concrete, in *Proc. SHRP-Conf.*, Götheborg, Sweden, 1989, VTI-rapport 352 A of the Swedish Road and Traffic Research Institute, 1990, pp 117–119.

44. D. G. Manning, Electrochemical Removal of Chloride Ions from Concrete, in *Proc. Symp. Elektrochemische Schutzverfahren für Stahlbetonbauwerke, Electrochemical Protection Measures for Reinforced Concrete Structures*, SIA Documentation D 065, Schweizer Ingenieur- und Architektenverein, Zurich, 1990, pp.61–68 (in German).

45. I. Uherkovich, "Praktische Aspekte der elektrochemischen Chloridentfernung", "Practical Aspects of Electrochemical Chloride Removal", *Proc. Symp. "Elektrochemische Schutzverfahren für Stahlbetonbauwerke", "Electrochemical Protection Measures for Reinforced Concrete Structures"*, SIA Documentation D 065, Schweizer Ingenieur- und Architektenverein, Zurich, 1990, pp.71–76 (in German).

46. M. Molina, "Erfahrungen mit der elektrochemischen Chloridentfernung an einem Stahlbetonbauwerk: Wirkungsweise und Beurteilung", "Experiences with the Electrochemical Chloride Removal at a Reinforced Concrete Structure: Working Principle and Assessment", *Proc. Symp. "Elektrochemische Schutzverfahren für Stahlbetonbauwerke", "Electrochemical Protection Measures for Reinforced Concrete Structures"*, SIA Documentation D 065, Schweizer Ingenieur- und Architektenverein, Zurich, 1990, pp.77–82 (in German).

47. H. R. Eichert, B. Wittke and K. Rose, "Elektrochemischer Chloridentzug", "Electrochemical Chloride Extraction", *Beton*, 1992, pp.209–213.

48. R. Polder and H. van den Hondel, Electrochemical realkalisation and chloride removal of concrete – State of the art, laboratory and field experiments, in *Proc. RILEM Int. Conf. on Rehabilitation of Concrete Structures*, Melbourne, Australia, 1992, pp.135–147.

49. R. Polder and R. Walker, "Chloride Removal from a Reinforced Concrete Quay Wall – Laboratory tests", TNO Report 93-BT-R1114, Delft, The Netherlands, 1993, 21.

50. B. Elsener, M. Molina and H. Böhni, Electrochemical removal of chlorides from reinforced concrete structures, *Corros. Sci.*, 1993, **35**, (5–8), 1563–1570.

51. J. Tritthart, K. Petterson and B. Sørensen, Electrochemical removal of chloride from hardened cement paste, *Ceme. Concr. Res.*, 1993, **23**, 1095–1104.

52. I. L. H. Hansson and C. M. Hansson, Electrochemical extraction of chlorides from concrete – Part I. A qualitative model of the process, *Cem. Concr. Res.*, 1993, **23**, 1141–1152.

53. R. Polder, "Chloride removal of reinforced concrete prisms after 16 years sea water

exposure", COST 509 annual report 1993 of project NL-2; TNO Report 94-BT-RO462, Delft, The Netherlands, 1993, 22.

54. R. Polder, R. Walker and C. L. Page, Electrochemical desalination of cores from a reinforced concrete coastal structure, *Mag. Concr. Res.*, 1995, **47**, (173), 321–327.

55. W. K. Kaltenegger and G. Martischnig, New gentle method of concrete repair, *Proc. 15th Slovak. Conf. on Concrete Structures*, Bratislava, 1994, pp.353–361.

56. K. Rose, "Elektrochemischer Chloridentzug - gleichmäßige Extraktion auch bei Bauwerksinhomogenitäten", "Electrochemical chloride extraction–uniform extraction also in the case of structural inhomogeneity", *Proc. Conf. "Beton-Instandsetzung '94", "Repair and Maintenance of Concrete and Reinforced Concrete Strutures"*, Institut für Baustofflehre und Materialprüfung, Universität Innsbruck, 1994, pp.53–59 (in German).

57. S. Chatterji, Simultaneous chloride removal and realkalisation of old concrete structures, *Cem. Concr. Res.*, 1994, **24**, 1051–1054.

58. J. Tritthart, Changes in the composition of pore solution and solids during electrochemical chloride removal in contaminated concrete, *Proc. 2nd CANMET/ACI Inter. Symp. on Advances in Concrete Technology*, USA, SP 154-8, 1995, pp.127–143.

59. B. Elsener and M. Molina, "Elektrochemische Chloridentfernung an Stahlbetonbauwerken", "Electrochemical Chloride Removal for Reinforced Concrete Structures", Vereinigung Schweizerischer Strassenfachleute, Zurich, VSS Report No. 507,1990 (in German).

60. B. Elsener, M. Molina and H. Böhni, Electrochemical removal of chlorides from reinforced concrete structures, *Mater. Sci. and Restoration*, Ed. F. H. Wittmann, Expert Verlag Ehningen, 420, 1992, 1, 792–804.

61. COST 509 "Corrosion and Protection of Metals in Contact with Concrete", Final Report, COST Secretariat, Brussels, 1996.

62. J. Bennett, *CORROSION '90*, Paper 316, NACE, Houston, Tx, 1990.

63. J. Tritthard, Chloride binding in cement — I. Investigations to determine the composition of pore water in hardened cement, *Cem. Concr. Res.*, 1989, **19**, 586–594.

64. J. Tritthard, Chloride binding in cement — II. The influence of the hydroxide concentration in the pore solution of hardened cement paste on chloride binding, *Cem. Concr. Res.*, 1989, **19**, 683–691.

65. R. B. Polder, Electrochemical chloride removal from reinforced concrete prisms containing chloride penetrated from sea water, *Constr. Build. Mater.*, 1996, **10**, (1), 83–88.

66. B. T. J. Stoop and R. B. Polder, Redistribution of chloride after electrochemical chloride removal from reinforced concrete prisms, in *Corrosion of Reinforcement in Concrete Construction*, Eds C. L. Page, P. B. Bamforth and J. W. Figg, Royal Society of Chemistry, Cambridge, 1996, pp.456–465.

67. R. B. Polder and J. A. Larbi, "Investigation of concrete exposed to North Sea water submersion for 16 years", CUR report 96-4, Gouda, The Netherlands, 1996.

68. R. B. Polder and J. A. Larbi, Sixteen years at sea, *Concrete*, July/August, 1996, 8–11.

69. R. B. Polder, Chloride diffusion and resistivity testing of five concrete mixes for marine environment, *Proc. RILEM Int. Workshop on Chloride Penetration into Concrete*, 1995, Eds L.-O. Nilsson and P. Ollivier, RILEM, 1997.

70. R. B. Polder and M. B. G. Ketelaars, Electrical resistance of blast furnace slag cement and ordinary portland cement concretes, *Proc. Int. Conf. on Blended Cements in Construction*, Ed. R. N. Swamy, Publ. Elsevier, 1991, pp.401–415.

71. B. Elsener and H. Böhni, Elektrochemische Chloridentfernung an Stahlbetonbauwerken", "Electrochemical Chloride Removal for Reinforced Concrete Structures", SIA Documentation D 099, Schweizer Ingenieur- und Architektenverein, Zurich, 1993, pp.161–163 (in German).

72. B. Elsener and H. Böhni, Electrochemical chloride removal – Field test, in *Proc. Int. Conf. on Corrosion and Corrosion Protection of Steel in Concrete*, Ed. R. N. Swamy, Sheffield, 1994, pp.1451–1462.

73. N. M. Ihekwaba, B. B. Hope and C. M. Hansson, Pull-out and bond degradation of steel rebars in ECE concrete, *Cem. Concr. Res.*, 1996, **26**, 267–282.

74. Ø. Vennesland and E. P. Humstad, Electrochemical removal of chlorides from concrete – Effect on bond strength and removal efficiency, in *Corrosion of Reinforcement in Concrete Construction*, Eds C. L. Page, P. B. Bamforth and J. W. Figg, Royal Society of Chemistry, Cambridge, 1996, pp.448–455.

75. O. Gautefall, E. P. Humstad and Ø. Vennesland, "Electrochemical removal of chlorides from concrete – chloride removal efficiency and bond stress", SINTEF Report, No. STF70 A 94103, 1995, 1–24.

76. G. E. Nustad, Desalination – a review of research on possible changes in the steel-to-concrete bond strength, in *Proc. Int. Conf. on Repair of Concrete Structures*, Svolvaer, Norway, 28–30 May 1997, pp.309–318.

77. J. P. Broomfield, Electrochemical chloride migration – Laboratory and field studies in the UK, in *CORROSION '97*, Paper 253, NACE, Houston, Tx, 1997.

78. W. K. Green and K. W. J. Treadaway, Electrochemical rehabilitation of concrete: Chloride extraction and realkalisation, COMETT Course, *The Corrosion of Steel in Concrete*, 17–19 February 1992, RWTH, Aachen, Germany.

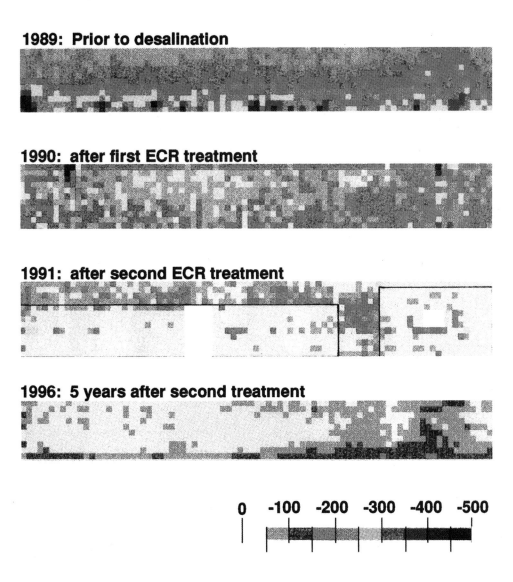

1989: Prior to desalination

1990: after first ECR treatment

1991: after second ECR treatment

1996: 5 years after second treatment

0   -100   -200   -300   -400   -500

Fig. 18 Potential field measured on a sidewall before chloride removal, after first and second treatments and six years after the treatments. Potential values are related to a $Cu/CuSO_4$ reference electrode [14].